LASER

THE INVENTOR,

THE NOBEL LAUREATE,

AND THE THIRTY-YEAR PATENT WAR

NICK TAYLOR

SIMON & SCHUSTER

New York   London   Toronto   Sydney   Singapore

SIMON & SCHUSTER
Rockefeller Center
1230 Avenue of the Americas
New York, NY 10020

Library of Congress Cataloging-in-Publication Data
Taylor, Nick.
Laser : the inventor, the Nobel laureate, and the thirty-year patent war / Nick Taylor.
p. cm.
1. Gould, Gordon, 1920–. 2. Lasers—History. 3. Physicists—United States—Biography. I. Title
QC16.G63 T39 2000
621.36'6'092—dc21
[B]   00--58803
ISBN 0-684-83515-0

For Richard Craven

LASER

1

"Call your first witness."

"Your Honor, I call Gordon Gould."

Gould pushed back his chair and walked toward the witness stand. The courtroom was not as large as some of the others he had seen. It was modern and low-ceilinged, with blond wood trim and polyester fabric on the chair seats, hardly grand enough, he thought, for the impact the trial would have, one way or another, on his life. But it would do, if only it were the last courtroom he ever had to see.

Six federal court jurors and two alternates watched Gould make his way to the enclosed area beside the judge's bench. They saw a man in his late sixties, sixty-seven to be precise. He wore a two-piece charcoal gray suit—that or blue, his lawyer had told him, no brown, no vests, and for God's sake none of those little thin-soled loafers that look like dancing shoes, wear lace-up shoes so the jury knows you're serious. It made him look serious, all right. Professorial. His high forehead topped by wavy gray hair and heavy dark-rimmed glasses hinted at the inventor his lawyer had described in his opening statement. "A very great inventor," he had said. A handsome man, too, it might be added, with a strong square face, a prominent nose, dimpled chin, and blue eyes that sparkled with humor. For all the buttoned-down sobriety of his attire, he looked like a man who had fun, a man you could like. He stood in the witness box and spoke the oath in a quiet voice roughened by a smoker's rasp.

"Do you promise to tell the whole truth and nothing but the truth, so help you God?"

"I do."

Gould found the chair behind him and sat down. The three or four spectator rows were loosely filled. Here in Orlando, retirees with time on their hands
jostled for seats at a good murder trial, crowding in in their loud shirts and the
women with pale blue sweaters for the air conditioning and everybody with
newspapers folded to the crossword puzzle for when the testimony got a little
slow. None of them were here. But a couple of reporters had their pens poised
to make notes. Lawyers representing firms like the one Gould was suing studied him, considering his potential effect on their employers. Shadow jurors
were sprinkled here and there, people hired to gauge the way the trial was going. Gould caught the eye of his companion, Marilyn Appel, seated in the second row. She gave him a smile of support.

At the defense table, expensively dressed, sat Robby van Roijen, wearing
his trademark aviator sunglasses. Van Roijen was the latest in a long line of
people to think they could sweep Gould under a rug and forget about him. But
he kept crawling out and popping up again. It was uncanny, but Gould was
persistent. To his left, U.S. District Court Judge Patricia C. Fawsett, fortyish
and blond, a nice contrast with the somber robes, presided. She was the third
judge in the ten-year-old case. All the paper in all the cases and other matters
that focused on Gould—the depositions, the transcripts, the Patent Office
pleadings, the files compiled on him by the FBI and branches of the military—
had accounted for the killing of a hundred forests.

Gould's lawyer turned his way. Warren Goodrich was a courtly man with
gleaming hair of snowy white and a voice like honey, perfumed with the kind
of homespun, folksy talk that juries loved. Gould had seen many lawyers, too.
Once he had counted, and stopped at around two hundred. Goodrich was a
trial litigator and something of a legend in courtrooms around Florida, including Orlando, where the Control Laser Corporation had its base of operations.
The patent lawyers who represented Gould had hired Goodrich for his facility
with jury trials.

He had opened with a ringing statement designed to impress the jury with
the importance of their jobs. "Ladies and gentlemen," he had said, "this case
involves an amazing, magnificent invention of man. It involves a machine that
has changed our lives. It has changed our industry, and changed our lives. A
laser."

Gould invented the laser. Sat bolt upright in bed one night in November
1957, grabbed a notebook, and wrote down the essence of the "amazing, magnificent" machine that, as Goodrich said, had changed industry and lives as
well as medicine and science. Then he went out and got his notebook nota-

rized. Those were the last smart things he had done for quite a while. So here he sat thirty years later in what he hoped was the last of many courtrooms, hoping for a verdict that would force laser makers to pay him for the right to manufacture his invention. Which they were reluctant to do, and that was putting it mildly.

The stakes were high. Gould figured to win or lose millions, depending on the outcome of the trial. It was just about the money, at least that was what he said. But in his heart, he knew a victory could mean vindication, too.

Quite a few people refused to believe Gould had invented the laser. The thinking was, he couldn't have. To do something like that, you had to be a serious scientist, or at least be an acolyte to one. Poor Gordon Gould, he didn't even have a Ph.D. And acolyte had never been a good way to describe him. So you were stuck with the insupportable notion that a thirty-seven-year-old graduate student had grasped the key to one of the more significant inventions of all time. Meanwhile, working down the hall was a very serious scientist and professor who, by the way, obtained the first patent on a laser and went on to win a Nobel Prize—before somebody figured out that the laser in his patent wouldn't work. Gould couldn't count the times his own claims had been greeted by the sound of laughter. If they wanted to be nice, those serious scientists, they would let on that Gould was clever enough but just, well, not really serious; he doesn't publish in our journals, you see, he doesn't go to our meetings. If they didn't want to be nice—and this was usually the case—they would hint that Gould had stolen his ideas from the august professor down the hall and then tried to claim them as his own. Hardly anyone but Gould was willing to say that it might have been the other way around.

Gould settled in his chair, thinking that if he won this thing, it would not only fill his pockets and bring his long ordeal to an end, it would force the keepers of the grail to rewrite their scientific histories. He wanted that more than he was willing to admit.

■■

The snowy-haired Goodrich stood near the railing of the witness stand and asked Gould to tell the jury a little bit about himself.

Gould leaned toward the microphone. He described his early life, his studies, his early attempts to make inventions, his return to graduate studies to learn physics when he realized he didn't know enough. He spoke quietly, matter-of-factly, until he began describing that November night in 1957.

Now something stirred. Gould, reliving the moment of his discovery, transformed in the witness box. He shed years of age and the lines in his face softened. His eyes shone and his voice crackled with new energy.

"It suddenly flashed into my mind that the amplification of light was going to do something quite remarkable, or could be made to do something remarkable," he said. "Instead of light going in every direction, you could generate a straight beam and concentrate the amplified light power into a beam of great intensity. You could focus that beam down to a tiny spot and get a fantastic intensity. And that thought was exciting. That immediately opened up my mind. My God, you could do things with this which have never been done before!"

Gould's prediction had been borne out in spades. There was almost no counting the uses that had been devised for lasers since Gould's original insight. He owed the fact that he could still see to one of them. People sped through supermarket checkout lines, used computers with virtually unlimited storage, communicated via cables that carried their voices by the thousands along beams of light, listened to music and watched movies in different ways because of lasers. Measurements were easier to make, and more precise. Manufacturers welded and drilled holes with lasers. Presidents and generals dreamed of them as both swords and shields.

It was hard to imagine that at the beginning, hardly anyone but Gould thought of the laser as more than an interesting scientific tool.

At the beginning. And how many beginnings there were to Gould's story. What did you want from it? The thrill of discovery was just a tiny part. Gould's tangled history was a mother lode of courtroom drama, professional backstabbing, government machination, corporate Goliaths marching on undermanned competitors, communists, conspiracy theories, romance, farce. But no matter how you cut it, it always came back to Gould's first wish, that he could imagine and produce new and useful things.

Close behind was the need to do it on his own, to work outside the box, for Gould was an outsider.

His hero was Thomas Edison. As a boy, Gould had reveled in accounts of Edison's journey from "addled" schoolboy to America's greatest inventor. Edison had seen things others couldn't. He had showed them, Edison had, and

Gould had found a delicious quality in the comeuppance of the great man's early critics.

Edison's special genius, Gould always thought, was that Edison invented things that people needed. His inventions transformed society and individual lives. As much as anyone, he had ushered in the modern age. But what Gould also found compelling about Edison was that he had mastered the *business* of invention. He had gotten his inventions into the marketplace and reaped their benefits. That meant he had conquered not only the mysteries of the science of his day, but an even greater mystery—the U.S. patent system. Edison had 1,093 patents to his credit; nobody had more. For this, Gould had the greatest admiration.

But that had been later. First, he had had to learn the tools by which he would invent.

▌▐

The lure of the sciences was his mother's genetic gift to Gould. In a family that on his father's side could be traced to the second voyage of the Mayflower, it was big-boned Helen Vaughn Rue Gould, a descendant of a French pirate, who bought her oldest son an Erector set and taught him to see the parts of things and not just the whole. She would sit with him on the oval hooked rug in the living room of their home in Pittsburgh, helping as he added and removed the pulleys, gears, and levers that were Gordon's first lessons in creative intuition. Kenneth Gould, an editor at his alma mater, the University of Pittsburgh, could deconstruct a sentence, parse a poem, and explain a social movement, but nuts and bolts defied him.

Kenneth left the university and joined a new magazine, *Scholastic,* in 1926, and when *Scholastic* transferred its operations from Pittsburgh to New York City in 1932, the Goulds moved with it. They took up residence in affluent Scarsdale, in Westchester County just north of the city, where they lived in a big house at 1 Berkeley Road.

Gordon was twelve, and just beginning to discover his affinity. He liked taking things apart, and soon the neighbors were bringing him their radios to fix. He attended Scarsdale Junior High and went on to Scarsdale High School, where he was a science whiz and, with his brother David, a member of the science club. David, eighteen months younger, had a flair for chemistry. Gordon was the more creative of the two, but David kept his nose in the books while Gordon was easily diverted by projects and adventures. Their mother would tell people that if only she could roll Gordon's inventiveness into one with

David's "straight A stick-to-it-iveness," she would have something like a perfect son. Geoffrey, six years Gordon's junior, was the only one of the three boys who looked like, and followed the interests of, their full-faced, stocky, and journalistic father.

Music was Gordon's other fundamental interest. He sang in the choir of St. James the Less Episcopal Church, and played the flute in the high school marching band and orchestra. He also belonged to the tumbling and chess clubs and, loving the outdoors, was a Boy Scout patrol leader and informal sailing teacher.

A portion of the boys' education occurred around the family dinner table. The world was in turmoil outside their comfortable enclave—belligerent fascism was on the rise in Europe; at home the Depression persisted, unions organized, and mine bosses, auto makers, and steel producers responded with police dogs, guns, and billy clubs, while an anti-communist fervor swept political life at all levels, including attacks on art and other forms of free expression. The senior Goulds discussed these developments at great length. They were liberals of the stripe known as "Norman Thomas socialists" after the liberal reformer who began a string of unsuccessful presidential runs in 1928. Kenneth was writing a book, *Windows on the World,* that was a high school level look at the systems of government then competing for dominance. He was treating democracy, communism, fascism, socialism, and cooperatives, ending with a grand flourish entitled "Man the Unconquerable," which was a progressive call to arms stating that spiritual values are more important than material ones, that the good life must leave room for free science, art, and religion, and asserting that humanity would survive "the poisons of power." A sense of utopian possibilities ran through the family's discussions. They may have been an intellectual exercise for Kenneth, but Helen had walked the walk, having taught at a remote settlement school in the Appalachian Mountains of Kentucky for several years before her marriage. In Westchester County, she was active in Democratic party politics and the League of Women Voters. By the time Gordon was a high school senior, he held progressive reforms aimed at achieving political justice as an article of faith.

One thing he never understood—the legend under his picture in the high school yearbook was, "This big, bold man." To his own eyes he was neither particularly big nor particularly bold. Gould, in high school and for years thereafter, had no grasp whatever of the appeal people, and women in particular, found in his martini-dry wit, his shock of light brown wavy hair, and the way his eyes laughed behind his small round rimless glasses.

Gould approached his high school graduation in 1937 ranked thirteenth in his class, with a choice to make. The Massachusetts Institute of Technology had offered him a scholarship. But MIT was out of state, and that meant that another, larger scholarship, one he'd received from New York State for being one of the top students in his class, couldn't be applied. His second choice was Union College, in Schenectady, New York. Kenneth Gould made a salary at *Scholastic* that kept his family in the middle class, but there wasn't much to spare. Gordon's choice of MIT already had been printed in the high school yearbook, but feeling the financial pressure, he switched grudgingly to Union.

■■

Schenectady straddles the Mohawk River above Albany. General Electric had its central offices and plant there, and stories circulated of how the great electrical engineer Charles Steinmetz liked to think and work on his ideas for alternating current while he floated down the Mohawk in a canoe, smoking thick, odoriferous cigars. He would think his way to the Hudson River and downstream to Albany, where he would debark to be shuttled, along with his canoe, by a company chauffeur back to Schenectady. Union fed the graduates of its electrical engineering department to GE. Physics was a stepchild of that department, existing mainly to give the engineers a basic grounding. When Gould entered Union, he was one of just four physics majors at the school.

Gould prospered at Union initially. He did well academically, joined the Sigma Chi fraternity, and went out for the swim team and track. He also belonged to the Outing Club, a group that went on excursions to the surrounding lakes and foothills. He worked as a waiter and at other jobs, and earned about half of the fees and expenses that his scholarship didn't cover.

He was not politically active, but he sought out political discussion. His family's dinner table talks had introduced the concepts of downtrodden minorities and the authorities, such as autocratic governments and greedy corporations, that exploited them. Gould had a strong sympathy for what his parents called "the working man," and believed that it was every person's duty to do what he could to make the world a better place. Now and then, prompted by these feelings for the underdog, he would talk late into the night with members of the Young Communist League. They talked about the perfectability of social systems, the obligation of governments to provide for all their citizens, and of companies to treat their workers fairly. The Young Communists tried to recruit Gould, but while he liked the discussions as intellectual exercise he didn't want to join the group. The members were too intense and humorless to be enjoyable. His closest friends at college were in Sigma Chi.

Union's Debating Society gave Gould another outlet. He was an aggressive debater, and wore his position on his sleeve if he was assigned the liberal position in a debate, say, on the march of fascism in Europe.

In his junior year, for the first time in his life, he confronted academic failure. For reasons that escaped him, he found rising in time to attend an eight o'clock mathematics class impossible and his teacher failed him. Fortunately for Gould, the secretary to Union president Dixon Ryan Fox became aware of his lassitude. She recognized the symptoms of mild clinical depression and, acting as a counselor, literally talked him out of it. He repeated—and passed—the course the following summer.

Physics, however, aroused Gould's interest without fail. Among the physics faculty was an expert in light sources named Frank J. Studer. Studer, a native of Michigan with a Ph.D. from the University of Wisconsin, was active in the American Physical Society and the American Association for the Advancement of Science. He was a lively speaker. In December 1940, when he delivered the physics department's annual Christmas lecture, he drew standing-room-only crowds to his experiments demonstrating what makes light and color. His presentation provoked so many questions he finally had to excuse himself for supper so he could return later and perform again for another eager audience of Schenectadians and high school physics students and their teachers.

Gould was as eager as any of them. Studer had awakened in Gould something elemental—a love for the beauty of light. Quickly, he grasped the science of optics as a window to that beauty. At Studer's hand, Gould first used prisms to separate white light into the spectrum of colors of which it is composed. Then, working in Studer's laboratory, he began to perform the kinds of experiments that had gradually revealed to scientists the dual nature of light—that it is sometimes a matter of waves, and sometimes of particles. He did two-slit and Frenell bi-prism diffraction experiments that showed the bands of light and shade caused when light waves interfered with one another. He bombarded atoms with electrons to produce red light from neon, blue from mercury, and yellow from sodium, showing light, as Studer sometimes put it, as "the violent shaking of atoms." At night, he haunted the physics building, frequently staying up in all-night sessions to photograph in long exposures the patterns light waves made passing around corners. Gould found that he could visualize the moving waves. He could picture them in his mind's eye. It was as if he could look at the aurora borealis and see not only its shimmering surface but the bombardment of the atmosphere by electrons from the sun that causes the phenomenon.

"It seems to me," Gould wrote in a letter home, "that all optical phenomena are beautiful."

■■

Gould was Union's sole senior physics major when he earned his bachelor of science in 1941. He then took his first job in his chosen field—a summer position at Western Electric in New Jersey. It was not at all what the eager graduate expected. He enjoyed his assignments in the laboratory. But among the executives, he noticed something striking. At each level of the hierarchy, the men—they were all men—occupied slightly larger offices than the last, the titles stenciled on their doors were slightly longer, and the hair on their heads was progressively more gray. The gradations could have been taken from a color chart, and each executive was infallibly polite to his superiors.

Gould realized he was looking at the corporate ladder. It seemed to stretch over his head into a gray infinity, and he didn't want to climb it. Gould wanted a different kind of life. He would work on his own, he decided, and invent things. Useful things that would improve people's lives, and from which he could make money.

Yale University's physics department at the time had a reputation for spectroscopy—the study of the spectra produced by the elements under various conditions—and it was to Yale that Gould went in the fall of 1941 to begin studies in optics and spectroscopy toward his Ph.D. His brother David was already there, beginning his final undergraduate year in chemistry. Gould lived in the Hall of Graduate Studies, sang in the Bach Cantata Club, and played the flute in a woodwind quintet that performed a baroque repertory. David played clarinet. The group was practicing Bach one Sunday in December when a student burst into the room where they were playing. Gould never knew his name, but he would never forget the mask of drama and dismay the young man wore as he blurted out the news that the Japanese had bombed Pearl Harbor.

Gould earned his master's the following spring, 1942. Educating scientists was important to the war effort, and he continued at Yale, working as an instructor while he continued to study for his Ph.D. But in late 1943, his job was cut to part-time status. This was not enough to maintain his deferment—the United States had almost every healthy man of eligible age in uniform—and

Gould was called for a draft physical. He passed easily, and made a perfect score on the science section of the screening test.

He was diverted before his induction date, however. Dr. W. W. Watson, the head of Yale's physics department, suggested that he visit a New York City address and apply for a job. "You'll be working in the war effort," Watson said. "And chances are you won't have to go into the Army."

In March 1944, Gould followed Watson's directions to Broadway and 137th Street in Manhattan. There he entered a huge, square, squat building that had housed the Nash Motors sales offices in New York. It still had ramps along the walls for moving cars from floor to floor. Gould met with the recruiting officer and filled out an application. It was accepted, and on April 30 he started work at what later became known as the Manhattan Project.

Gould's job at the Manhattan Project had been a young scientist's dream, but out of it had grown a long nightmare. Gould still wasn't exactly sure how it had happened. Sex, pride, anger, young rebelliousness—all had played a role.

He had been twenty-three, full of energy, enjoying his work. Each day, he rode the subway uptown from Greenwich Village, where he had taken a small apartment at 204½ West Thirteenth Street, just west of Seventh Avenue. In that one subway ride he traveled from bohemia to America at war.

There was an exhilarating sense of urgency about it, the breathless race against the Germans to unlock the secrets of the A-bomb that would defeat the enemy. And each time his idealistic conscience tapped him on the shoulder to remind him that he was helping to build a weapon that could wipe out humanity, he told his conscience that atomic power would offer humanity a plentiful supply of energy once the war was done.

The code name for their work was the S.A.M. Project. The initials stood for Special Alloys Material, which was supposed to explain why a building full of scientists and technicians were working on aspects of uranium. At the beginning, everybody knew everybody else and there had been no security precautions. William A. Nierenberg, a tall, uncommonly bright Columbia University graduate student who headed Gould's section, remembered that as the project grew, somebody brought a camera from home and took pictures that went into badges that everybody wore. By the time Gould arrived, the Federal Bureau of Investigation was handling security, and everybody had to fill out forms.

Later, when the project fell under the U.S. Army Corps of Engineers, Army Intelligence took over, and things got really serious.

In the laboratories of the Nash Building, Gould joined a number of people building a pilot plant to test the isotope separation cascade being built in Oak Ridge, Tennessee. The only way to produce bomb-grade uranium, it was thought at the time, was by passing uranium hexafluoride gas through a series of membranes to separate its isotopes from one another. This was an exacting process involving literally hundreds of miles of airtight piping divided into some four thousand membrane-separated stages, each of which isolated a fraction more of the precious U-235 necessary for the bomb. A day's production would be a teacup's worth, but eventually there would be enough.

The model on which Gould worked was not nearly so elaborate as the real thing. It had only twelve stages, and with its pipes and valves and U-turns looked like an industrial-sized model of the digestive system. It was designed to test the membranes and the various other components of the real apparatus, especially the vacuum seals. Gould's title was research assistant. According to his paycheck, he worked for Columbia University, the New York City institution where work on the project had begun.

One day in June, about four months after he had started, Gould heard a sharp, earthy, female laugh. He looked up from the calculations he was doing on a slide rule to see a group of young women being shown through the model cascade. Between radar development and the rush to beat Hitler to the bomb, the war had absorbed most of the country's twenty thousand physicists and practically everybody else with advanced technical training. Still, there were jobs to do, and the Armed Services Training Program was recruiting and teaching nontechnical people to do them.

"There were maybe six of them, and they were all attractive," Bill Nierenberg said, recalling the stir the young women caused in the laboratory. "Their job was to find and seal leaks in the piping of the model cascade. They wheeled around carts, with these helium leak detectors. They'd spend days in the lab, going over the equipment, trying to find all the leaks."

Gould soon fixed his attention on one of the new arrivals. She was tart-tongued, with brunette curls and a vivacious smile, although what first stood out to him was her hip-swinging walk. Glen Lincoln Fulwider had attended Sweet Briar College in Virginia. She should have graduated in the class of 1942, but she'd left sometime before then and come to New York, where she worked for a time at the American Optical Company.

Up to then, Gould's romantic life had been straitlaced. He had had some moments at Yale with a woman named Ruth, but as a serious matter he was dating a woman he met at a Quaker peace conference the summer he graduated from high school. Her name was Caroline Gerber. Caroline had gone off to London as a volunteer at the beginning of the war. Now she was back in New York, living with her rich parents on Park Avenue while trying to become a dancer. Caroline was a catch, and Gould loved her. On the other hand, she had a lot of plans. She was pushing to get married, promising to work while he finished his Ph.D., eager to start a family once he had it. He started to see her as a domestic version of the corporate ladder, and despite his feelings for her, he ended the relationship.

Gould found he could talk to Glen and forget about Caroline. They talked deep into the night, conversations that touched on the arts, politics, the war, and social problems, and Gould would not know where the time went. He learned that her swingy walk was the result of hardened tissues caused by tuberculosis of the hip. They talked about the world's full quotient of injustice. Glen wasn't pushing Gould in the direction of the altar, and certainly did not consider marriage a prerequisite to sex.

Glen's frankness about sex was an allure to Gould. The pirate in his mother's background was said to have snatched his future wife from a beach in England and spirited her to the new world, but the lustiness in the gene pool had been tempered by a more recent run of dour rectitude. Gould's two grandfathers were Methodist ministers, one cerebral, the other, his mother's father, a circuit-riding preacher who moved from town to town. His aunt Margaret on his mother's side was a missionary who had spread the gospel in China to people she would never stop calling "the heathen Chinee." Gould had it drummed into him that sex created obligations. In Glen's view, sex was its own reward. Into this oasis of zesty and unburdened coupling walked a man who was parched.

Soon Gould was deep in a wild, bed-shaking romance, sweaty in the summer's heat. His downstairs neighbors would plead for quiet by rapping the ceiling with a broom handle. He and Glen would laugh and loll in bed and gaze out at the stained glass windows of the faux Gothic church across Thirteenth Street. If the pounding bothered them too much, they moved a stone's throw away to her apartment, in a building just on the other side of Seventh Avenue.

Uptown at the Nash Building, they worked at a furious pace. They dripped sweat in the un-airconditioned laboratories, and swallowed salt tablets to ward off dehydration.

■■

Almost as alluring to Gould as Glen's sexual attitudes was her political outlook. She was fiery in her concern for exploited people. Gould could hardly believe their views meshed so completely. Glen actually was a step or two beyond the Gould family's genteel liberalism. After she had left her parents' home, during what she called her "starvation days" in Boston, she had been taken in by a family of Italian socialists, and embraced their politics. Gould thought she was thrillingly progressive.

The fall of 1944 brought news from Europe of the tide of war turning in favor of the Allies. After landing at Normandy and pushing through France, the American and British armies hovered near the Rhine. The Red Army had pushed the Germans west nearly to Budapest and Warsaw. Gould, reading the newspapers and listening to the radio, was caught up in these joint successes. Fascism was on the run, and America's brave Soviet allies were giving a good account of themselves.

Glen was even more enthusiastic. She could hardly contain her admiration for the Soviets. One day, she suggested that Gould join her at a meeting of an informal Marxist study group in Greenwich Village. There were many such groups around New York City then, organized around talk of current events or, more formally, around political agendas. The war and the competing systems of democracy, fascism, and communism had their counterparts in the salons and classrooms of the ever-vocal and argumentative city. Glen said a friend had told her about the group, she had gone and found it interesting, and she thought he would, too. He agreed, and one fall night they walked downtown to an apartment building on Downing Street, just off Sixth Avenue. Headlines at the newsstands that they passed trumpeted an American naval victory in the Pacific's Leyte Gulf.

When they got where they were going, they joined half a dozen other people in the living room of a small one-bedroom walkup. There was a fervor in the air that reminded Gould of the Young Communists back at Union. After a few minutes of awkward conversation, accompanied by bargain bin wine and yellow cheese on crackers, a thin, diminutive, soberly dressed and somewhat threadbare man got up to lead the discussion.

The little man proved a knowledgeable and energetic teacher. His eastern European accent seemed to Gould to convey centuries of cultivation, and he waved his arms like an orchestra conductor when he spoke. His command of Marxist theory was encyclopedic. The group discussion quickly became a dialogue between Gould and the leader, whose name was Joseph Prenski.

Prenski was in his middle thirties. He had immigrated to the United States from Poland and studied social sciences at Syracuse University, where he received a bachelor of arts in 1932. After working toward a doctorate, but failing to win a fellowship that would sustain him, he left the university and came to New York City, where to support himself he taught at the left-wing Jefferson School of Social Science on Sixth Avenue at Sixteenth Street, and at private study groups around the city. Some of the private groups were organized by the Communist Political Association, the name chosen by the Communist Party in the United States in an attempt to recast itself and its public image that same year. Prenski had joined the party a year earlier, in 1943, but left that off his resumé.

Gould returned with Glen to the class a week later. Soon they were regulars, reading Marx on political economy and Lenin's treatises on the state and revolution and imperialism as capitalism's inevitable spawn. The group discussed the Marx-inspired reforms of the Soviet Union and their effectiveness, industrial production, agrarian reform, central economic planning, and the equality of workers. Gould was routinely the most vocal questioner. He enjoyed the back-and-forth with the animated, intellectually preening Prenski. They were clearly the two smartest people in the group. The others included carpenters, secretaries, and commercial artists, whom Gould regarded as members of the proletariat.

Gould's romance with communism started out as intellectual. He had been a political tenderfoot, romantic and naive. As his father would have said, he "knew a lot that ain't so." He assumed Soviet-style communism was as advertised, that it was providing widespread benefits across social and economic lines, and that it was supported by popular consensus. And out on the Eastern Front in Europe, brave Russian boys were whipping the Nazis. Hitler was the common enemy, and the Soviets and Americans were fighting him together.

Gould had been attending the study group with Glen for several weeks when Prenski suggested that he join the Communist Political Association. It would be a good way to translate rhetoric to practical activity, Prenski said. Gould ignored the suggestion. Socializing with communists, as he had learned at Union, could be an almighty bore, and he preferred to choose his friends from among the physicists and other scientists he worked with. He enjoyed the theoretical discussions, but he wasn't a convert.

▮▮

The new year, 1945, had barely begun when word circulated among the employees at the S.A.M. Project that it was being taken over from Columbia by the Carbide and Carbon Corporation, which would become known as Union Carbide. Concerned workers traded rumors about their new employer. They had heard that the company had a reputation for discrimination against Jews. Word circulated that while Jewish employees were welcome to stay on the project, they could not expect to advance within the company. Gould and Glen both were vocal in protest. Why fight Hitler if we were going to allow discrimination on the home front? Gould asked indignantly.

Word filtered down that employees should keep their opinions on nonscientific matters to themselves. This was exactly the behavior Gould expected from indifferent, faceless corporations that cared only about profits, and it fanned the flames of his idealism. Ignoring the warnings, he continued to protest.

Soon afterward, Army Intelligence took over security duty from the FBI. FBI agents had been satisfied to keep a low profile while they observed employees of whom they were suspicious; the Army's G-2 agents brought down the ax.

A letter arrived at the Nash Building dated January 20, 1945. "Gentlemen," it said, "Investigation conducted on Richard Gordon Gould, who is employed by your company as a Research Assistant, reveals information of such a nature as to make undesirable his employment on any work of interest to the Manhattan Engineer District.

"It is therefore requested that his employment be terminated immediately and that this office be advised of such action.

"Under no circumstances is this employee to be informed that this release is at the request of the Manhattan Engineer District, Military Intelligence Service, or any other military or civilian agency."

At noon the following Friday, January 26, Gould was summoned from the laboratory and dismissed with two weeks' notice. Glen was called from her helium leak detector and told that she was fired, also with two weeks' notice. They were among those who at least had a chance to leave with a modicum of dignity. Others were fired on the spot. Bill Nierenberg watched as one secretary was escorted out of the building so abruptly a policeman had to come up from the street to retrieve her purse from her desk. None of those summarily dismissed were given reasons for their firing.

Gould walked out of the Nash Building two weeks later a changed man. He had loved his job, was proud of it, and it had been taken from him. He was angry at the authorities, and humiliated in the eyes of his family and friends. The purge had confirmed his worst suspicions. It illustrated the communist argument that private enterprise was a source of injustice that could be alleviated only by government action. Not knowing what to tell his family, he said he and Glen had been fired for protesting Union Carbide's anti-Semitic policies.

It would have saved him a world of trouble, he thought later, if he had known the true reason.

4

The tides of fortune that were sucking Gould down were lifting him up at the same time. He didn't know it, but Bill Nierenberg had been looking out for him as he worked out his two weeks. The lanky Nierenberg was smart in ways well beyond science. He was only a year older than Gould, a son of immigrants who was born on New York's Lower East Side, barely removed from the tenements. He had studied at the Sorbonne before the war, and knew the sympathies that fueled leftist thought in the intellectual circles of the day. Much of it was simply fashion, like the Chicago post-debutantes he had met in Paris who lived their lives according to what they read in *The New Yorker*. He regretted losing Gould, because Gould was a hard worker and Nierenberg didn't think he was a spy. So when a professor from New York University showed up looking for somebody who knew something about vacuum systems, Nierenberg called Gould in and introduced them.

Serge Alexander Korff was a youngish, sandy-haired associate professor of physics. He was moonlighting as a consultant to a New York mirror and glass company that was struggling to fulfill a military contract to produce large front-surface mirrors.

Korff explained the problems the company was having. Gould thought he could handle them with one hand tied behind his back. Korff had said, apologetically, that the job was not permanent, and probably not full-time. That was fine with Gould, because an invention had come to him and he wanted time to work on it. He agreed to take the job.

■■

Semon Bache, the company that Korff had referred to, was on Greenwich Street in the western part of Greenwich Village, two blocks from the teeming Hudson River steamship piers. It was an easy walk from Gould's apartment.

In the Village streets, shopkeepers and pushcart operators, tradespeople and laborers, working women and housewives out shopping, all seemed united in a common expectation. The western Allies had repelled the German counteroffensive in the Battle of the Bulge, the Red Army had pushed through Poland, and American naval forces continued to gain in the Pacific. The air was full of hope for a quick end to the war.

The glass company's building took up the entire block between Morton and Barrow Streets, fronting on Greenwich, with loading docks on Washington Street. It contained administrative offices, manufacturing areas, and a warehouse.

Korff met Gould at the door and introduced him to several company executives and members of the board. Then they took him to a room where the faulty mirrors were made. It contained a vacuum tank that was an oversized version of a standard laboratory bell jar. It looked like a Jules Verne conception of a diving bell, a massive metal dome six feet from edge to edge and four feet high, studded with rivets and suspended by a crane over a circle in a raised steel platform, beneath which were vacuum pumps to draw out the air. Rising from the platform was a tree of electric filament coils.

A group of the mirrors was being readied when Gould and his escorts arrived on the scene. As a burly foreman barked orders, men in coveralls arrayed three-foot squares of glass on metal easels around the tree of filaments. There were three sets of the tungsten filaments, each with cups that held an element vital to the mirror-making processes. The first held powdered chromium, which when heated to a vapor would coat the glass with an adherent surface. Then aluminum powder, also heated to a vapor, would add the shiny reflective surface. Finally, silicon monoxide would add a clear coat to protect the shiny aluminum and keep it from growing dull from oxidation. Properly made, a front-surface mirror provides a better image than an uncoated glass mirror since the layer of slightly distorting glass over the reflective surface is eliminated.

When the glass sheets were in place, the workers added measured amounts of the necessary elements. The foreman issued more orders, and a young man about Gould's age pulled a lever. The dome descended, and the workers positioned it on the circular groove that contained a rubber O-ring that would make the vacuum seal. Another switch set the vacuum pumps to work. It took the better part of an hour to void the tank of oxygen. While everyone waited, one of the board members regaled Gould with tales about the foreman, "one of our Puerto Rican brothers," the man said. He hadn't gotten the hang of the mirrors, but the management loved him because he once had strapped on a pair of revolvers and threatened the workers to head off a rumored wildcat strike. When instruments indicated that the vacuum finally was achieved, the man at the control panel flipped more switches to activate the filaments. They heated quickly. Thick glass windows allowed a limited view inside the dome. Peering, Gould saw them glowing white hot at 1,500 degrees centigrade, as the materials on them evaporated and dispersed onto the mirrors.

A few minutes later, the dome creaked skyward revealing mirrors that Gould knew at a glance wouldn't pass muster at a carnival funhouse. They were wavy, blotchy, and in some cases, transparent.

"What are they doing wrong?" Korff asked him.

Gould told him that the key lay in precisely calculating the placement of the filaments in relation to the mirrors so that the elements coated the glass evenly.

"Can you do it?"

"I think so," Gould said.

"Good." Korff introduced him to the foreman. The man narrowed his eyes at Gould's youth, his glasses, the slide rule case dangling from his belt, his pocket full of pens, and the notebook he carried to record his calculations. He narrowed them further when the board member mentioned that Gould had gone to Yale and that the company expected the foreman to follow Gould's directions.

Gould had returned the next day and started taking measurements. The foreman had observed him with contempt. He stood close and crooned, "What you doing, college boy?"

Gould explained the process.

"So, how long you been making mirrors, college boy?" the foreman said. "As long as me, you think?" The strikebreaker clearly was no friend of the working man. The benevolent socialism Gould envisioned did not anticipate such harshness between classes.

"When you going to make the perfect mirrors, college boy?" the foreman wondered.

"Give me a couple of weeks," Gould said.

There were thirty-six filaments on the tree centered in the vacuum tank. Gould worked in the shadow of the bell suspended from the crane, measuring the positions of all the filaments. A covering lay on the floor to preserve the cleanliness of the vacuum chamber. When the chromium, aluminum, and silicon monoxide reached evaporation temperatures, the atoms or molecules would fly toward the mirrors. Since they flew out evenly in all directions, Gould could calculate geometrically the relative density of the layers they would form at a given distance from the filaments. The challenge was to arrange the filaments to eliminate overlaps and thin spots. They were easy enough to adjust with a few twists of a wrench, but the coatings were so thin, about one tenth of a micron, that the calculations for their placement had to be just right.

Gould finished his measurements and went home. Two weeks later, he returned to the workshop with a set of diagrams. When the filaments were adjusted to their optimum positions, he checked them. "I think we're ready to try it now," he said.

"So, the college boy is ready to make perfect mirrors," the foreman sneered.

Korff hurried over from the nearby NYU campus. Soon the contract manager and other executives gathered. The foreman gave his orders and the shop workers mounted the sheets of glass on their easels. Gould and the control operator measured the coatings in their powdered form into the filament cups. The dome descended and nestled onto its O-ring. Gould stood by nervously waiting for the vacuum pumps to do their work. Then the relatively quick process of heating the filaments began. Minutes later, the crane hoisted the dome away.

The group crowded forward to look at the results, and even the presence of the executives could not stop the foreman from leading the way. Gould was sweating; the mirrors looked good, but he couldn't tell without a close inspection. Nobody said a word as the group walked in a circle, inspecting each of the mirrors in turn. At last the board member who had advertised him as a Yalie said, "By God, they look perfect."

The foreman circled again, peering closely at the mirrors, as Gould was pulled away by the board member. "Thank you, my boy," he was saying as he pumped Gould's hand up and down. "Nobody could figure this out until you got here."

Then Gould was surrounded by company executives offering congratulations as the foreman shouted at the shop workers to ready a new round of mirrors for production.

■■

Once he had proven himself, Gould's duties at Semon Bache lapsed to those of a paid consultant. He immediately turned to his invention—an improvement to the recently developed contact lens. Gould had worn glasses since he was five years old, and yearned for the freedom to pursue the outdoor activities he loved, especially his favorite water sports, without having to choose between the inconvenience of glasses and blurred vision.

Contact lenses had been around since before the turn of the century, but only in the 1930s did glass and hard plastic contact lenses start to be relatively widely used for cosmetic reasons as well as vision correction. Before 1942 a blue-eyed motion picture actor, Walter Hampden, was fitted with a pair of brown contact lenses so he could convincingly play the part of an American Indian chief. The convenience of going without spectacles was offset by discomfort and poor fit, because the lenses covered a large portion of the eye. Even after developments in plastics allowed contacts to be made from molds taken from patients' eyes, the hard lenses did not allow the cornea to breath, and they could not be worn for more than a few hours. Most wearers found they had to remove the contacts during the day to let their eyes recover. But if inserting the lenses was troublesome, removing them was worse. One device developed for the task was a small rubber suction cup mounted on a stem. Its users looked as if they were stabbing themselves in the eye with an inverted mushroom cap.

Gould knew that delivering oxygen to the eye, and removing the carbon dioxide "exhaled" by the cornea, would allow contacts to be wearable for longer periods. He filled notebooks with sketch after sketch of his ideas.

Working on his own, Gould had the freedom he had always wanted. But he quickly found that puzzling over an invention was no substitute for the scientific collegiality he had known at the Manhattan Project and even briefly at Semon Bache. It had been a part of his life that was very much an anchor. His spirited dialogues with Joseph Prenski notwithstanding, his social life and his most invigorating contacts had grown out of his scientific associations.

Now Glen was his only soulmate and companion. They both were victims of the corporate bosses who didn't like them standing up for Jews, and their common grievance drew them closer.

Soon after they were fired, Glen announced that she had joined the Sixth Avenue Club, a local branch of the Communist Political Association. She urged him to attend a meeting with her, and he did. Soon afterward, he joined the club himself. Members told him he'd be smart to use an alias. He chose the name John Thacker, and received a membership card and a dues receipt in that name.

The Sixth Avenue Club had plans for a bright young man like Gould. Flattered by the club's attention, he agreed to become its literature director. He assembled Marxist pamphlets and books that the club sold to its members. He stood on Village street corners, hawking *The Socialist Worker* and passing out flyers decrying the plight of exploited workers. Some people spit at him, others said they hoped he burned in hell. Gould told himself he was doing missionary work aimed at pushing America toward social justice and equality.

His brother David, fresh out of Yale with his doctorate in chemistry, moved into the Thirteenth Street apartment, and Gould slept at Glen's apartment. She joined him on family visits home to Scarsdale, and they double-dated with David and his girlfriend, Toni. Glen was ripe for a political discussion regardless of the company, although discussion was not quite the word to use when one talked politics with Glen. She was not just firm, but strident in her beliefs. Gould's family began to wonder why in the world he put up with her.

His family's developing distaste for Glen, and his parents' concern about his defense of Marxist doctrine, drove him further into her camp. Gould was enamored of his own humanity, his great concern for the workers. He liked the shock his opinions caused his parents. Visits home ended in pitched arguments.

In the city, Gould attended the 1945 May Day parade and passed out pamphlets. Later the same month, while America was celebrating Germany's surrender, the Communist Political Association deposed Earl Browder as its national president and elected firebrand William Z. Foster, under whom the Communist Party U.S.A. reemerged from the shroud of the so-called "association."

Three months after the war in Europe ended, the atomic bombs dropped on Hiroshima and Nagasaki ended Japan's resistance to surrender.

The war's end brought waves of returning servicemen, ready to start families and renew their lives. New York City landlords responded with a campaign to end the price and rent controls imposed by the federal Office of Price Administration. Gould found in the threat of rent increases a new issue to drive his leftist sympathies, and he enthusiastically joined the Sixth Avenue Club's efforts to organize rent protests among apartment dwellers.

About this time Joseph Prenski suggested to Gould that he join a professional club of the Communist Political Association. The idea was that he would have more in common with lawyers and accountants than the carpenters, commercial artists, and secretaries who belonged to the Sixth Avenue Club. This he did. There were no dues and no card to carry. Nor was there much purpose to the club as far as Gould could tell, other than keeping up to date on party activities and encouraging other professionals to join. Gould joined a series of discussions, held at the homes of members, about economics and political theory, but found them dreary. He preferred the workings of the Sixth Avenue Club, where at least there were rallies and literature handouts to prepare for.

Gould's flurry of pamphlet-pushing and protest organizing still left time for work. He turned out new sketches of his contact lens idea, but remained unsatisfied. Another invention had occurred to him, this after he had learned that Serge Korff had a moribund vacuum chamber in his laboratory at NYU that he was willing to let Gould use if Gould could bring it back to life. He also was still working for Semon Bache, and visited the mirror company to remeasure his settings each time the filaments were cleaned.

■■

By early 1946, Gould had devised a contact lens that used the eyelid's natural blinking action to pump oxygen under the edges of the lens. The same blinking action removed carbon dioxide "exhaled" by the cornea. Pumping tears through and under the lens and across the cornea would allow the eye to breathe, and wearers would experience less of the discomfort and clouding that many complained of.

Gould was confident his drawings could be turned into an effective, even revolutionary, contact lens. But now, he realized to his dismay, he had no idea what to do with them.

His research on contacts had included a 1942 book, *Contact Lenses*. Its author, Theo E. Obrig, headed Obrig Laboratories, a small Manhattan-based maker of molded plastic contacts. He tracked down Obrig to his East Forty-eighth Street office just off Fifth Avenue, and asked for an appointment. But when he sat down with the lens maker, Gould was seized with paranoia that his invention would be stolen.

Obrig was curious about what the bright young man was after, and waited expectantly to be informed. Gould started, stopped, and stammered, but with the details of his invention on his tongue, he couldn't get them out. He was afraid he was going to reveal too much, and that Obrig could make the lens improvements on his own. He had no business experience, no idea how to protect himself, and no basis for suggesting any mutually satisfactory arrangement such as a joint venture.

With Gould unable to talk, Obrig was unable to respond. Gould left the office gripped by frustration, a Marxist craving business acumen.

In later years, Gould would prove his naivete and lack of sophistication in combining business and invention over and over again.

By the end of 1945, Gould had left the Thirteenth Street apartment and moved west to Bank Street, where he was joined by his second brother. Geoffrey had signed onto an oil tanker as a purser at the end of the war. A few months at sea satisfied that impulse, and he left the ship when it returned to New York and moved in with Gordon while he decided his next move. In their spare time, the brothers amused themselves flying kites from the roof of the building, Gordon insisting on playing out twine until he calculated the kites were a mile away, flying over New Jersey.

Gordon's politics continued to strain the family ties. David and Toni had moved to an apartment on East Ninetieth Street after their June marriage. One night, when Gordon visited, they got into a pitched argument in which Gordon took a Marxist line. The argument went on until three in the morning, and Gordon left without budging from his radical views.

The aftertaste of the argument was so bitter that Gordon never discussed politics with his brother and sister-in-law again. He did, however, ruin the family's Thanksgiving dinner in 1946, arguing for Marxism so strenuously that he and his father almost came to blows.

Helen made things worse. She tried to intervene, telling Gordon his political views were going to ruin his career. That made him even angrier, and he and Glen left after exhausting and embarrassing the entire family. He felt more than ever that the two of them, as outcasts, had to present a united front.

While he defended socialist and Marxist programs, an instinct for self-preservation kept him from admitting his ties to the Communist Party. This was reinforced when, in the summer of 1946, he took an instructor's job teaching physics in evening sessions at the City College of New York.

By now, Gould was spending less time on communist activities. Semon Bache had asked him to resume a full-time schedule, and the evening teaching was the equivalent of another full-time job. What little time he had left over he preferred to spend on experiments and research. He maintained his ties to the clubs, however, even joining the Teachers Union at the professional club's suggestion. The clearly left-wing union represented teachers in the New York City schools and instructors in the city colleges.

■■

When the family gathered in Scarsdale for the Easter weekend in 1947, Gordon cleared his throat at the end of the meal and said, "I have something to announce."

Everyone laughed, for it was an ill-kept secret that he and Glen had revived their once-postponed wedding plans. In the weeks leading up to the wedding, Glen's friends threw a party to which Gordon's family was invited. About forty mostly young people attended, and there was dancing. It was obviously a communist function, and Glen was treated as a big shot. Gould's father, who attended without Helen, talked about his support for the New Deal with one of the other guests but resisted his companion's insistence that this meant he was a socialist in everything but name. The communists had tried to dupe him once before, in 1936, when he had agreed to serve as an editorial advisor to a magazine called *Champion of Youth*. His advice was never called upon, and not until the early 1940s, when the magazine and the names on its masthead came up in hearings of the House Un-American Activities Committee, did Kenneth learn that *Champion* was a publication of the Young Communist League. Now that anti-communism was a more powerful force than ever in American politics, he was alert to anything that could damage his name and reputation.

Gould and Glen were married at the Scarsdale house on June 7, with her friend Louisa DiMassimo, the daughter of the Italian socialists she had stayed with in Boston, as Glen's bridesmaid. The next day, they left on a honeymoon that consisted of a driving trip through New England. Their first night's lodging was a picnic table at Bear Mountain State Park, where a park ranger awakened them before dawn to tell them the area was closed at night.

It seemed to his family during this time that Gordon was politically stunted around Glen. He either took her side, or clammed up. Glen, on the other hand, never hesitated to speak. Helen and Kenneth had seen the famous husband-and-wife acting team of Alfred Lunt and Lynn Fontanne in the Robert Sherwood play "There Shall Be No Night" soon after Gordon and Glen were married. They were discussing the play, a critical look at the Soviet invasion of Finland, when Glen looked up from her detective novel and said, "Do we have to talk about this hogwash?"

But tensions developed quickly in the marriage, for by then Gould's romance with communism was fading. Teaching at City College had renewed his contacts with other scientists, broadening his social life beyond Glen and her friends and relieving the sense of embattlement and isolation he had felt with her after they were fired. At the same time, the Soviet Union was spreading its influence ruthlessly over eastern Europe, and Gould was appalled at what he saw. He had always thought Czechoslovakia was a grand little country, a democratic island in a sea of despots, struggling to do the things his own vision of liberal reforms encompassed. President Edvard Beneš's postwar government had been pro-Soviet but independent. When a communist coup toppled the elected government early in 1948 and forced Beneš, one of the country's founders, to resign, Gould's disillusionment deepened. The Soviet blockade of West Berlin that June made it complete.

▌▌

During this time Gould was working feverishly on a new invention. He had decided that he could grow diamonds in a vacuum chamber, and had completed repairs to Serge Korff's machine to begin his experiments.

Korff's chamber was a miniature version of the vacuum tank Semon Bache used to make its mirrors. It was a standard laboratory tool that was an upended glass jar a foot and a half in diameter and two feet high, set on a metal base, with an electric heating coil running through it. Grease at the juncture of the glass and the metal base preserved the vacuum seal.

The reason Gould thought he could use it to make diamonds lay in the atomic structure of a diamond. To the naked eye, and even under a normal microscope, the surface of even an uncut diamond appears unbroken. However, at the atomic level, it presents a lattice that offers a toehold to similar atoms introduced in the right way. Gould's plan was to heat carbon in the form of graphite in a vacuum chamber and evaporate it onto a small diamond. He thought the carbon atoms, in a frenzy of movement produced by the heat, would move around until they found their niche in the crystal lattice. In theory, they would adhere to the diamond, and enlarge it.

He tried the experiment a hundred different ways, varying the heat and other conditions. The process was laborious and slow—pumping the atmosphere inside the chamber down to a vacuum, then heating a tungsten cup containing the graphite, and also heating a rough but clean host diamond about an eighth of an inch in diameter. Night after night, all he got were smudges of graphite.

He discussed the problem with his brother David, who told him the thermodynamics of diamonds were such that it would not grow. Gordon stubbornly kept trying. After several dozen more failures, he decided his brother was right.

With that, Gould realized that if he was ever going to invent anything useful and valuable, he needed to know more about physics. In the summer of 1949, he applied to enter the graduate program in physics at Columbia.

Columbia was one of the premier physics institutions in the country, but Gould had chosen it for convenience as much as anything. The campus between Broadway and Amsterdam Avenue in an uptown neighborhood called Morningside Heights was just two stops south of City College on the Broadway IRT subway line, a twenty-one block ride to or from his teaching job. From the stop at 116th Street and Broadway, you climbed the stairs to daylight, and entered the campus through a massive iron gate.

A brick walkway led from the gates to the campus center, a vast open quadrangle dominated by the domed and columned Low Memorial Library that sat at the top of a set of wide steps at the quadrangle's north edge. Gould's passage was to climb the steps, skirt the left side of the library and head toward the northwest corner of the campus and the Pupin Physics Laboratories, passing through a series of small courts that the architects McKim, Mead and White had designed into the campus until he reached Pupin, a somewhat homely building thirteen stories tall.

Charles Hard Townes occupied a ninth floor office in Pupin. Five years older than Gould, at thirty-four Townes was one of physics department chairman I. I. Rabi's prize recruits. He was a tall Southerner with an aristocratic bearing who was beginning his second year on the Columbia physics faculty. Rabi had lured him away from the Bell Telephone Laboratories the year before.

Everyone who knew Townes considered him brilliant. He was from Greenville, South Carolina, but after graduating from high school at sixteen and earning undergraduate degrees at Furman University and a master's in physics at Duke, he went to CalTech for his Ph.D. and two years of postdoctoral work. Since coming to New York, he had proven to be a Renaissance man, taking courses in music and voice at the Juilliard School of Music.

Townes's eight years at Bell Labs had spanned the war and two years on either side. His work at the research laboratories had focused on airborne radar used in bombing controls. World War II Army Air Corps radar used radio waves 3 centimeters long, which oscillate at a frequency of 10,000 megahertz. The Army and its scientific advisors thought a shorter wavelength radar would paint a better radar picture; instead of a blob, it might show a plane, and even reveal its type. It would also be lighter and more compact. After the war, the Army asked Bell Labs to develop radar using a 1.25 centimeter wavelength. Townes predicted that this 24,000 megahertz radar would fail because the molecules in atmospheric water vapor typically absorbed waves of that frequency.

The Air Force—the Air Corps was made a separate service branch in 1947—ordered the system built anyway. When it failed as Townes had predicted, his early reputation was assured. He brought his interest in high-frequency radio waves, or microwaves, and their interaction with molecules, with him to Columbia, where he pioneered the new field of microwave spectroscopy and forged the beginnings of a remarkable career in science.

Gould had arrived at Columbia about the same time as a thickset, humorous jazz enthusiast and postdoctoral fellow named Arthur L. Schawlow. Schawlow, twenty-eight, had been born in Mount Vernon, New York, but his mother was a native of Canada and the family had moved to Toronto when he was three. He grew up and attended school there, and when World War II interrupted his doctoral studies he worked in a radar factory on microwave antenna development. He resumed his studies after the war, and after he received his Ph.D., won the fellowship that sent him to Columbia to work with Townes.

Townes and Schawlow were destined to be linked in ways that in retrospect seem likely. They had similar scientific interests, and Townes's forceful and persuasive personality fit well with Schawlow's more taciturn nature.

That their paths would entwine as they did with Gould's, however, would have defied the most fanciful imagination at the moment when Gould, unknown to them both, took a deep breath and walked into Pupin.

■■

Columbia at the time was rebuilding a physics department whose talent had been scattered by wartime assignments. The department's reputation was prodigious, and its history virtually inseparable from that of physics in the United States.

Columbia University in the City of New York, as it is officially named, did not confer a Ph.D. in physics until 1897, over a hundred and fifty years after its founding in 1754 as Kings College, the sixth institution of higher learning chartered in what would become the United States. Robert A. Millikan received that Ph.D. Millikan went on to win a Nobel Prize in physics, and the department's tone was set.

The American Physical Society, a sort of domestic Royal Academy for physicists, was founded at Columbia in 1899. Soon after the turn of the century, $50,000 worth of bonds donated in memory of a young alumnus, Ernest Kempton Adams, by his father allowed department chairman George B. Pegram to invite distinguished scientists to Columbia. H. A. Lorentz, Wilhelm Wien, and Max Planck, all Nobel winners whose work contributed to the evolution of relativity and the quantum theory of atomic and subatomic particles, came to Columbia as Ernest Kempton Adams lecturers. Pegram tried to get Albert Einstein, too, but Einstein, who by then had authored the theory of special relativity, wrote back in January 1912 that he was too busy to come. Pegram was successful, however, in bringing Niels Bohr to Columbia for a series of lectures in 1923, the year after Bohr won the Nobel Prize.

The department's modern reputation began to take shape in 1936 when Pegram, by then dean of the Graduate School, dangled an appointment before the Italian physicist Enrico Fermi. Fermi waited until 1938. Then, when he went to Stockholm, Sweden, with his family to receive that year's Nobel Prize in physics, he left Mussolini's Italy behind, continued on to New York, and started teaching at Columbia the following year. Fermi's work, and that of Harold Urey, who discovered heavy water and won the Nobel Prize in chemistry, laid the foundations for the Manhattan Project.

After Fermi's atomic pile and research on sustained nuclear reaction were transferred to Chicago, the Columbia Radiation Laboratory was involved in the war effort through the development of magnetrons and klystrons for powering radar.

When the war was over, the job of restocking the physics faculty with brilliant and demanding scientists went to Isidor Isaac Rabi.

Rabi was born in Galicia in what is now Poland and came to the United States as an infant. He grew up in New York's Lower East Side tenements and later in Brooklyn, studied electrical engineering and chemistry at Cornell, but switched to physics and earned his Ph.D. in physics at Columbia while teaching at City College. Two years of postgraduate work in Denmark and Germany between the wars brought him in contact with pioneers of quantum theory, including Bohr and Werner Heisenberg, who would win the Nobel for physics in 1932. Returning to Columbia in 1929, he taught and did research in the years leading to World War II. The Manhattan Project attempted to recruit him, but Rabi thought radar would be a more immediate help to the war effort, and took a leave of absence to serve as associate director of the MIT Radiation Laboratory in Boston.

When he returned, in 1945, he was Columbia's most recent Nobel Prize winner, and the first to have done his Nobel work at Columbia. Rabi in 1937 had invented an improvement to molecular and atomic beam machines that were used as experimental tools for studying the elements. In their basic form, they produced beams of molecules or atoms by heating to a vapor, in a furnace, the element to be studied and releasing the vapor through a series of slits into a vacuum. Researchers applied magnetic fields to the beams and studied the way the atoms and molecules reacted. Rabi's device took the process one step further, passing the beams through a magnetic field and then exciting the resulting transitions with pulses of radiation. This allowed him to record the magnetic properties of atomic nuclei, for which the Nobel committee awarded him the prize in 1944.

Rabi was short, about five-two, but in no other way was he diminutive. The department in which Gould was about to begin studying bore the stamp of his intellect, his specialty, and his forceful, overbearing style.

7

Gould met the limits of his knowledge quickly at Columbia. The physics he'd absorbed up to that time was classical and rudimentary. Quantum mechanics was the next step, but the courses he'd taken in that rarified discipline at Union and at Yale had succeeded only in mystifying him.

Fortunately, as it turned out, Columbia would not accept the credits for the quantum mechanics course he'd taken in Yale's graduate program, and he was obligated to repeat it. The course at Columbia was taught by Rabi.

Quantum theory states that undisturbed atoms can have only certain discrete amounts of energy. Take the extremely simple hydrogen atom, with a nucleus of a single proton orbited by a single electron. When it is excited from the ground, or normal, state by the right amount of energy—from heat, say, or electricity—the electron jumps to one of its known higher orbits and nowhere in between. When it relaxes, returning to the lower state, it does so by the same amount. Picture a bull's-eye with rings that are varying distances apart. Each ring from the center corresponds to a higher energy level, and the energies between these levels are never observed. So when an atom's energy is absorbed or emitted, it is always in discrete amounts, or packets, called quanta that correspond to the distances between the rings. The movements between energy levels are called transitions. In the world of atomic and subatomic particles, quanta are entering and departing all the time though an atom's tendency is to remain in, or return to, its ground state. An atom may absorb energy in the form of photons of light or other radiation, or when it collides with another atom or electron and energy is transferred between them. In a neon sign, for example, electrons are driven down a tube exciting neon atoms as they go. Classical physics could not account for such phenomena. It took quantum mechanics and its understanding of electromagnetic radiation as consisting of both waves and particles to explain them.

Rabi had learned quantum mechanics from Heisenberg, who along with Erwin Schrödinger and others was developing the emerging set of principles and their mathematical framework when Rabi studied with him in Germany. It was like learning perspective in art from Leonardo, and Rabi had the gift of imparting the knowledge with the same seminal clarity.

Gould, hearing Rabi explain those principles and how they had evolved, learned for the first time how to think about quantum mechanics.

Gould's gift in physics was his intuition, a peculiar and quite special way of seeing physical processes at work. At Union, it had given him insight into the activity of light waves. Rabi's lectures now gave him an image of activity at the subatomic level. The difficulty quantum mechanics presents to most people is that the processes it describes happen at a scale vastly smaller than any ordinary experience, and so it is usually described in the terms of its mathematics. Rabi could think mathematically, using theories to predict behavior, and also experimentally, from the feel his experimental knowledge provided of the physical processes. Gould was no theoretician, but he could imagine the processes at work. He found he could think physically about the interaction of atoms and light. He could actually picture what was going on, and with the insights he gained from Rabi's lectures he was able to ponder new possibilities.

Added to his expertise in optics, quantum mechanics gave Gould the basis for profound advancements in his thinking.

■■

Outside the Pupin building's lecture halls, Gould's increasingly rocky life with Glen was tinged with melodrama.

Though his ardor for Soviet communism had cooled, he remained a committed liberal. He contributed to left-wing causes including the American Committee for Yugoslav Relief, the American Peace Crusade, American Youth for Democracy, the Guardians Club, the Jefferson School, where Prenski had taught, and the Veterans of the Abraham Lincoln Brigade. He also had given $100—it was supposed to be a loan—to the Civil Rights Congress for a bail fund for eleven U.S. Communist Party leaders on trial in New York charged with advocating the government's overthrow. Glen, of course, could always get excited about a liberal cause, the more celebrated the better, and

Gould had been just starting at Columbia when their sympathies took them into a near-fatal riot.

The occasion was a concert featuring Paul Robeson near the Hudson River town of Peekskill.

Robeson, the great black bass-baritone and stage actor, was a controversial figure. He spoke out against oppression, racial crimes, and segregation, and advocated friendship with the Soviet Union though he was not a communist. The week before, a racist mob had attacked a small group of would-be concertgoers at a picnic grounds outside Peekskill where he was scheduled to perform. Robeson refused to be intimidated, and agreed to return the following Sunday, which was Labor Day weekend. A coalition of three unions organized a guard detail. One of the unions was the Teachers Union, of which Gould had remained a member.

Gould and Glen rode a chartered bus to Peekskill, where they joined an audience of five thousand in a natural ampitheater on the grounds of the Hollow Brook Country Club. Two thousand union men stood shoulder-to-shoulder around them. This time, the mob waited until the concert was over, then attacked the concertgoers as they headed home. Rioters hurled stones at buses and cars, and dropped bricks and concrete blocks from overpasses as local deputies and New York State troopers did little to interfere. Gould and Glen huddled in the aisle as rocks smashed through the windows of their bus and showered the screaming passengers with glass.

The attack confirmed Gould's liberal instincts, but did nothing to redeem communism in his eyes. Glen, meanwhile, was increasingly shrill in its defense. His growing suspicion of Soviet aims and tactics was matched by her hardening position in support of anything the Soviets did. She could say with a straight face that Joseph Stalin was as great a man as Abraham Lincoln. She applauded when communist North Korea poured sixty thousand troops armed with Soviet guns and tanks over the border into South Korea on June 25, 1950.

Gould objected to Glen's enthusiasm. He said that as a scientist he had learned to reexamine his assumptions, and it was time for Glen to reexamine hers. She accused him of selling out the people.

Their arguments increased in frequency and volume, and by September 4, 1950, Gould had had enough. Pressed by the demands of teaching at CCNY and his course work at Columbia, and tired of the shouting, Gould moved out of the apartment at Thirteenth Street and Seventh Avenue and left Glen to her politics.

■■

Charles Townes, in the meantime, advanced in 1950 from associate to full professor. The same year, he was named executive director of the Columbia Radiation Laboratory. His career had assumed the steep trajectory it would maintain for years.

In 1951, Arthur Schawlow completed his two-year fellowship with Townes. By then, he had met and fallen in love with Townes's younger sister, Aurelia, and the two had married. Townes was slated to be the next physics department chairman. Anti-nepotism rules at Columbia prevented Townes from hiring his new brother-in-law, so he steered Schawlow to Bell Labs, where Townes still worked as a consultant. Schawlow was assigned to work on superconductivity, an area that interested him less than microwave spectroscopy, and in his spare time he continued to work with Townes.

■■

Gould finished two years of classroom work in 1951. He and Glen reunited briefly, found their diverging political views irreconcilable, and made their separation official. Gould moved to an apartment at 122nd Street and Riverside Drive, near Grant's Tomb and the faux Gothic Riverside Church. On a good day it was walking distance from both CCNY and Columbia.

Ready to begin his thesis work, Gould was accepted as a doctoral student by Professor Polykarp Kusch. Kusch was a big, meticulously dressed man who wore dark-rimmed glasses, combed his thick dark hair in a razor-sharp part, and smoked cigarettes using a short holder. He had a booming voice that could be heard two floors away. His office was next to Townes's, who was forced to add soundproofing in order to converse with his students.

Kusch, a German native whose family immigrated to the United States when he was one, had been at Columbia since 1937. He had worked closely with Rabi in molecular beam spectroscopy before World War II, and after a succession of wartime jobs working on aspects of radar, high-frequency oscillators, and microwave generators, returned to Columbia and his primary interest, spectroscopy.

Rabi's magnetic resonance method, for which he won the Nobel Prize, allowed formerly elusive properties of particles to be precisely measured. Kusch, confronting emerging questions about the equation in quantum mechanics that described known properties of an electron, was able to use magnetic resonance to make a new measurement. He found that an electron's intrinsic magnetic moment, due to its spin on its own axis, was slightly larger than the orbital magnetic moment caused when an electron revolves around a nucleus. This precise determination of the magnetic moment of the electron was pure science; it had no obvious application but expanded the understanding of quantum mechanics.

In the physics department, where Rabi's style and interests held sway even though he had rotated out of the chairmanship, professors were prone to put their thesis students to work solving such esoteric problems. There were hundreds of permutations of matter to be measured using molecular and atomic beam machines, and many incremental additions to the emerging understanding of the way particles interacted with electromagnetic radiation.

■■

A mid-century set designer looking to furnish a mad scientist's laboratory needed to look no further than a beam machine. They were complicated and cumbersome devices. Their guts were tubes of metal a foot in diameter, heavy enough to withstand severe pressures and high heat radiating from the furnaces that glowed with the extreme temperatures it took to vaporize substances from the table of elements. The tubes were about four feet long in the case of an atomic beam machine, and twice that for its molecular counterpart—this because a molecule's magnetic moment is weaker than that of an atom, and therefore takes longer to react as it passes through a magnetic field. The tubes were flanged on the ends and contained thick glass or quartz viewing ports. Tubes ran out of them, for sucking out the air, and wires ran in for powering the magnets that would bend the beams.

Surplus submarine batteries supplied the power. They were as big as the side of a house at eight feet tall, and added to the look of perverse experimentation. One could imagine the machines blinking, shuddering, and belching jets of steam.

They were complicated to operate, as well. This meant that a student just embarking on thesis work that involved beam experiments "apprenticed" to another who was further along. The newer student learned the machine while providing the other with needed experimental help, recording measurements and so forth.

Kusch assigned Gould to work with a student named Alan Berman, whose thesis project was a standard molecular beams experiment—measuring the hyperfine structure separation of the ground state of thallium, a metallic element that in its solid form resembles lead. To avoid distractions, Berman and Gould worked largely at night. Keeping the equipment running was a constant battle. They had to be alert to the beam machine's frequent vacuum leaks, and they also fought electrical noise and other disturbances as they sat in a darkened room straining their eyes at the swing of light reflected from a moving coil galvanometer.

Berman finished his thesis work in the fall of 1951. But instead of turning Gould loose on an experiment of his own, Kusch first asked Gould to work with another student. Peter Franken, like Berman, was younger than Gould, and precocious. He had been working on a universal atomic beam detector as his thesis experiment before Rabi, with Kusch's approval, ordered him switched to a less complicated experiment that Franken thought was beneath his skills.

Nevertheless, Gould worked with Franken for six months while they used the standard atomic beam machine to measure the hyperfine structure of the ground state of potassium. In the process the two became good friends.

Finally, at the end of 1952, Kusch gave Gould his thesis experiment assignment. He wanted Gould to build on Berman's work by measuring the energy separations in the hyperfine structure levels of thallium in its metastable state. Mathematically, this is the doublet P three half state, terms (and symbols) that provide measures of the characteristics of the shape and configuration of the atoms. It turned out to be a far more daunting task than either Berman or Franken had faced.

■■

Hyperfine structure separations are states quite close together in their energy. Measuring such separations in the ground state of elements, as Berman had done with thallium and Franken with potassium, was relatively easy because the ground state is the normal state, in which the vast majority of all atoms and molecules exist at any given time. They may absorb energy, from heat or light, for example, but in an instant they spontaneously emit that energy, and return to the ground state. Boltzmann's law governing the distribution of atoms postulates that the number diminishes sharply with each step up in the level of energy, or excitement; the greater the energy, the fewer the number of atoms or molecules that have that much energy. This is true even when an element is heated to a vapor.

The metastable state of thallium or any other element is the first state above the ground state. Unlike higher states, it has a long lifetime—a hundredth of a second, say, as opposed to a hundred millionth of a second. Such lifetimes fascinated Gould. He always had a sense of wonder when he thought about individual atoms and the speed at which they got from one place to another.

The problem facing Gould was not the measurement itself. Rather, it was getting enough of the atoms into the metastable state to measure. Gould thought he could use heat to excite them, and as 1953 began he set to work designing modifications to the atomic beam machine.

8

Gould's marriage to Glen was headed toward a merciful annulment, and he was dating steadily again. He had first met Ruth Frances Hill at Yale when they were students there during the war. She, like Gould, had a master's in physics when they met. They had enjoyed each other's company, in and out of bed, but had fallen out of touch. Meeting again at Columbia, where Ruth was working on a Ph.D. in biophysics, they resumed the relationship.

Gould once told a friend that his two great fears in life were public speaking, and good-looking women. Ruth, who was three years older, was not particularly attractive. Nevertheless, she was fun to be with after Glen's rigid ideology, which had exhausted and ultimately repelled him. They soon were partnering in weekend bridge games with friends. Weekends during the boating season found them on Long Island Sound in a small sailboat they kept at the Norwalk, Connecticut, home of Gordon's brother Geoffrey, who now worked as a reporter for the Associated Press in New York.

■■

At Pupin, new developments in the understanding of electromagnetic radiation were taking place. Charles Townes and two of his graduate students were working on a device that would foreshadow a newly defined field—quantum electronics. It would create a new means of studying elements through spectroscopy, and revolutionize thinking about an almost forgotten prediction of Einstein's. It had its genesis in 1951.

That spring, Townes and Art Schawlow were attending the American Physical Society's meeting in Washington, D.C. The brothers-in-law decided to save money by sharing a hotel room. It was Schawlow's habit to work, and sleep, late. Townes had young children and was accustomed to rising early. One morning, not wanting to disturb his brother-in-law, he went out for a walk. When he reached Franklin Park, he sat down on a bench to enjoy the weather.

Many interviews that have been done with Townes since then have referred to "the epiphany of Franklin Park," for it was then that Townes realized that a long-held tenet of thermodynamics did not necessarily apply. One consequence of the second law of thermodynamics is that atoms or molecules at a given temperature are limited in the amount of energy they can radiate. And since under Boltzmann's law more atoms or molecules will always be at lower than at higher energy states, much of the energy emitted by the higher states is always absorbed by the lower. But this assumes they are in thermal equilibrium. Townes suddenly realized you didn't have to have thermal equilibrium. You could use artificial means to separate the atoms in a higher state from those in the lower.

This would set the stage for the fulfillment of one of Einstein's predictions. He had predicted as early as 1918 that excited atoms and molecules would be stimulated by passing radiation to emit their energy as more radiation; that is, the energy of a wave would be increased by the energy of the emitted radiation. Thus wave energy could be amplified. By producing or isolating a large enough population of excited atoms or molecules and feeding it into a resonant cavity, you could form a standing, or a continuous, wave.

Back at Columbia, Townes began to suggest to some of the graduate students working under him that they consider trying to make such a molecular oscillator. They discussed the problem at great length in Townes's office. A studious and mild-mannered graduate student, James P. Gordon, and a postdoctoral fellow named Herbert J. Zeiger did some preliminary calculations and fiddled with designs. They found the prospect daunting. The project Townes was suggesting, according to their calculations, was just on the edge of possible. No graduate student wanted to waste time on a thesis experiment that was likely to fail.

Gordon, however, decided that even if the device didn't work as a self-sustaining oscillator, it could still work as a spectrometer for analyzing so far unseen hyperfine structure in the ammonia spectrum. That was enough to build a thesis on. He held his breath and took on the project as a thesis topic, and Zeiger worked with him.

After Gordon and Zeiger had been working for more than a year, Rabi and

Kusch walked into Townes's office. "Look, that's not going to work," Rabi announced. "You know it's not going to work, we know it's not going to work. You really ought to stop. You're wasting money, and you're wasting your graduate students' time."

Townes resisted the pressure, though he was thankful he had tenure. "No," he said, "I think it has a good chance of working. We'll keep on a little longer."

By then one of the biggest problems with the experiment was solved. In the rectangular metal box they had built in Townes's main laboratory room, the students had used strategically placed electrical fields to separate the excited molecules in an ammonia vapor from those in the ground state and focused them into a beam. In ammonia there was a fairly high percentage of such molecules, and they stayed excited for a relatively long time. The problem that remained was devising a cavity in which the excited molecules would play off one another and form a standing wave. Initially, they enclosed the cavity at its ends to keep microwaves from leaking out. But finally, Gordon decided to try opening the cavity's ends. The cavity redesign produced a standing wave of such great efficiency that energy leaking out was not a problem.

Early in April 1954, more than two years after he had started working on the experiment and Zeiger had moved on to a private laboratory, Gordon knocked on the door of the tenth floor room where Townes held a regular Friday afternoon microwave seminar at which students discussed their work, and exchanged suggestions. Heads turned toward the door as Gordon opened it and popped his head inside.

"It worked," he said.

The seminar broke up and headed to the lab. Townes's response, after the initial euphoria had dissipated, was to say, "Let's build another one." The device they had spent two years building could only be tested by its twin.

At first the boxlike creature had no name. Then Townes said it needed to be called something. He was adamant on one point only. "Nothing ending in -tron," he insisted. Eventually he, Gordon, and some of his other students settled on maser, an acronym for microwave amplification by stimulated emission of radiation.

After the experiment's success, Gordon, Zeiger, and Townes published the results in *Physical Review Letters*, a journal of the American Physical Society. When it appeared, the maser was big news, and not just in the scientific world, but in the popular press. *The Wall Street Journal* ran a front page story.

The excitement wasn't about the maser's applications. With their extremely stable frequencies, masers proved to be virtually noiseless amplifiers with uses as receivers in radio, astronomy, and highly accurate atomic clocks. But to the world at large, the maser wasn't particularly useful. Its most important function was never its few scientific applications. It was something larger, a step in science that opened the minds of researchers to new possibilities, by bearing out Einstein's prediction, and proving that electromagnetic radiation could be amplified and self-sustaining.

■■

Gould, struggling with the design of his modifications to the beam machine, cursed his cleverness in avoiding Army service during World War II. The G.I. Bill would have helped with his tuition payments and reduced his long hours of teaching. A tutor in physics at City College performed a kind of academic slave labor. The job paid according to the hours taught, offered no tenure, and required an annual reappointment.

Having started at $2,550 including a cost of living bonus in the 1946–47 academic year, Gould had received steady increases as had all the city's teachers. He earned $3,950 from August 1953 through July 1954, and special funds provided by the City of New York were to push the next year's salary to $4,400. He also taught for eleven weeks in the summer class of 1954 and earned $5 per hour.

The city's Board of Higher Education met on October 25, 1954, and approved a retroactive raise to $5.45 per hour. Gould's extra pay before taxes amounted to $57.60.

By then, he needed every penny.

Fear of communism had sustained a fever of witch-hunting in the United States. It was embodied nationally by the U.S. House of Representatives Committee on Un-American Activities and its jailing for contempt of the "Hollywood 10" writers and directors who refused to affirm or deny their affiliations. Senator Joseph McCarthy of Wisconsin upped the ante of paranoia with wild, unfounded charges of communists in the State Department and the U.S. military.

Anti-communist paranoia also gripped state legislatures. While nationally, public attention focused on Washington and Hollywood, the red-hunting fervor of the states was aimed at unions and academic institutions. In New York, attacks on alleged communists had long been a political staple. Between 1919 and 1940 there were three legislative attempts to purge "reds" from the public payroll, and the City College faculty was a regular target.

The last of these, the Rapp-Coudert Committee, named for its co-chairmen, was established by an act of the New York State legislature in 1940, and conducted investigations through 1941 and 1942. It was dissolved before Gould began lecturing in physics at City College in 1946. But the anti-communist fervor had not abated in the legislative halls. In 1949, state lawmakers passed the Feinberg Law, aimed at rooting out subversion in the public schools. It was challenged by the Teachers Union, and held unconstitutional.

Not to be denied, Albany passed the law again a year later, and this time, it was upheld. And in April 1953, after near-riots and picketing by public school teachers, and the suspension of teachers who refused to name names or answer questions about their party ties, the legislature amended Feinberg to apply to New York's municipal colleges.

The law required the state Board of Regents to promulgate rules. Under them, the city's Board of Higher Education listed the Communist Party as a subversive organization that advocated violent overthrow of the government. Party membership was "prima facie evidence of disqualification for appointment to or retention of any position in our colleges."

Prospective faculty members had to sign declarations that they were not and had never been members of the Communist Party or "any organization that advocates the violent overthrow of the Government of the United States or of the State of New York or any political subdivision thereof." If they had been, they had to say so, describe the circumstances, and prove that they had severed all connections "in good faith" before October 23, 1953. They also had to make "full and fair disclosure of all the facts" arising from their party connections, which meant they had to reveal the names of other members.

The board appointed a special committee to deal with Feinberg, which in turn formed a special investigating union headed by board special counsel Michael A. Castaldi. Castaldi started dragging City College faculty into closed meetings.

Gould was called to appear before the unit on May 21, 1954. The summons took him to a faceless building in lower Manhattan, where he went alone and entered a room in which three people sat at a table. Gould took a seat in front of the table. A legal stenotypist sat to one side, her fingers poised over a steno machine.

"Mr. Gould, were you ever a member of the Communist Party?" the questioning began.

Gould admitted his party membership. He said he had attended party meetings from 1946 to 1948, read communist classics as assigned reading, marched in May Day parades, and passed out pamphlets and the *Socialist Worker.* He said he had probably visited the Sacco-Vanzetti Communist Party Club headquarters in New York, had been sympathetic at the time to Marxism-Leninism, and had contributed to party projects including the bail fund. He described his disillusionment starting with the coup in Czechoslovakia in 1948, and departure from the party shortly afterward.

But when it came time to name names, Gould found he couldn't do it. He refused to say who had invited him to the first meeting, would not disclose the homes of members where meetings were held or the names of persons who attended, and he would not tell his interrogators anything about them. He kept mum about Glen, and the estrangement and divorce that followed their political divergence.

The meeting ended noncommittally, as far as Gould could tell. He felt he had made a good case for himself. He had been forthcoming and, he felt, convincing in his opposition to communism and Soviet values. And he had stood up for his own values in not revealing the secrets of others. He left with the impression that what had been said was private and sacred. Everything in his dossier was secret, and would remain within the Board of Higher Education. He walked out into a fine spring day, glad that the ordeal was over.

In August, a letter arrived in the mail informing Gould that his contract at City College would not be renewed.

The fall semester at Columbia was almost upon him. He had no job, no income, and his apartment lease—he had moved two years earlier from Riverside Drive to West 187th Street in Washington Heights—was up for renewal. He solved the apartment problem by moving in with Ruth, who lived on the ground floor of a small building her parents owned in the Bronx. They were spending almost all their time together anyway. But there was no sign that he was going to be able to solve his thesis problem anytime soon. He went to Kusch in despair and explained his predicament.

The professor grew visibly angry as Gould told his story, drumming his cigarette holder against the wooden desk where he sat with his back to the window. Kusch considered Gould a bright and promising student, and he wanted him to see his thesis through.

"I'm going to have to drop out," Gould said finally. "I don't see any other way."

"No, you won't drop out," Kusch boomed. "This is an outrageous situation. I'm going to see what I can do."

He persuaded Gould to stay on and keep working while he tried to arrange a paid research assistantship.

9

Gould remained at Columbia, living on Ruth's income as he continued to struggle with a way to excite enough thallium atoms into the metastable state to carry out his thesis assignment. What worked for Townes, Gordon, and Zeiger in the maser, diverting the ground state molecules from the excited ones in an ammonia vapor, would not work for Gould and his thallium.

All the elements have unique properties, and they comprise atoms and molecules that have their own sets of discrete energy levels as they ascend from the ground state. Ammonia's molecules above the ground state have energy levels that are closely spaced, and highly populated. Thallium in its metastable state has widely separated and higher energy levels, and they are almost totally unpopulated.

After months of experimentation Gould still was stumped.

He had tried to do it thermally, adding to the standard tungsten oven a tantalum superoven that would heat the thallium to 2,500 degrees centigrade, well beyond its evaporation temperature. That took six months, and failed to produce enough excited atoms for his measurement. He spent another six months bombarding the thallium beam with electrons. That, too, produced too few excited atoms to measure. Gould continued to think, to calculate, and to tinker in his laboratory on the ninth floor of Pupin, but he was increasingly frustrated at having no clear path to his thesis with a year already gone by.

He was working in the lab one day in 1955 when Rabi burst into the room. Gould respected Rabi enormously, and feared him not a little. He knew Rabi's tendency to demand that things be done his way, and also to bully the speak-ers at the regular public science colloquia held in the physics department's third floor amphitheater. Graduate students, postdoctoral fellows, and faculty members from schools like Harvard, MIT, and Johns Hopkins were invited to present lectures on promising work. Rabi would sit in the front row with Kusch and a few others, and hurl questions at the speakers. It took a strong ego, and a command of the subject, to withstand Rabi's inquisitions. But Gould remembered Rabi's lectures in the same steeply tiered room, which had re-vealed to him the anatomy of quantum mechanics.

The professor was excited. He said that he had just returned from a scien-tific meeting in Paris, where he had met the French scientist Alfred Kastler. Kastler, who taught and was co-director of the physics laboratory at the Ecole Normale Superieure, was developing optical methods to study certain reso-nances in atoms. It was an equivalent to what Rabi and Kusch had been doing for years with molecular beams, but Kastler was working at the much higher optical frequencies—the frequencies of electromagnetic radiation that form light waves.

Rabi explained that Kastler had developed a new technique he called opti-cal pumping. He was using resonance radiation—light caused by bombarding atoms with electrons and forcing them to absorb and spontaneously emit light energy—to "pump" other atoms of the same element from their ground state to higher energy levels at which they could be studied. This emission of reso-nance radiation takes place in an electrical discharge tube, like a neon sign.

"He's getting large numbers of atoms into excited states by optical pump-ing," Rabi said. "Why don't you try that?"

Gould immediately understood that Rabi had given him the solution to his problem. He started making sketches of a discharge lamp.

The lamp had to be made of quartz. Glass would melt at the 1,100 or 1,200 degrees centigrade needed to get a thallium vapor. He designed a quartz tube with electrodes through the ends. The thallium would be heated in the tube to evaporation temperatures, then high voltage applied across the electrodes would send electrons crashing into the thallium atoms. The electrons would knock off other electrons—this is the discharge for which such a lamp is named—and also rouse the thallium atoms into the higher states of excitement from which the characteristic green resonance radiation of thallium could be produced.

That light, falling upon the beam of atoms issuing from a slit in the oven of the atomic beam machine, would excite the atoms by providing photons of the right energy. Some of the excited atoms would fall back to the ground state, others to the metastable state. But the end result, Gould hoped, would be enough atoms in the metastable state all at the same time to do his measurements.

■■

A not unfamiliar swell of pride ran through Pupin in the fall of 1955 when the Nobel prizes were announced. The Swedish Royal Academy awarded Kusch a share of the Nobel in physics for determining the magnetic moment of the electron. He shared it with Willis Lamb, who had also done his Nobel work—analyzing the fine structure of the hydrogen spectrum—while at Columbia. Lamb had since left the faculty, but excitement and a renewed sense of specialness inhabited the physics department. Not since Rabi's award in 1944 had the Nobel gone to Columbia physicists for work done while they were there.

■■

Gould worked through 1955 and into 1956 at the design of his thallium discharge lamp. That April, Kusch finally managed to secure him a job as a part-time assistant in the Radiation Lab. It paid $200 a month, which didn't come close to matching what Gould had made at City College. But he was so used to being poor by then, and having to scrounge even the price of cigarettes from Ruth, that it seemed like a fortune.

They had gotten married on a Sunday in the fall of 1955 at a judge's chambers in White Plains. Ruth had her Ph.D. by then. She had earned it in February 1954, and gone on to postdoctoral research work at the Columbia-Presbyterian Medical Center, where she studied the effects of ultraviolet radiation on bacteriophage, or the viruses in sewer waste. She also, as she liked to say, did "an occasional stint of teaching" students in radiology. In her vision of her life with Gould, they both had doctorates, teaching appointments with tenure, led a comfortable academic life full of intellectual and cultural pursuits, and played bridge with friends in their spare time.

Gould liked bridge, but he didn't exactly share Ruth's vision. He still burned with the inventor's fire, and for now all his attention focused on the lamp he hoped would solve his problem.

Things kept breaking, and Gould would wait six weeks while the laboratory's glassblower fashioned a new lamp. Finally, the thing was ready. Gould's optically pumped atomic beam machine was not a thing of beauty. It was a kludge, a Rube Goldberg device that Gould would have to clean and repair every few days. But it worked.

The thallium atoms issued from the oven through a slit no wider than a knife edge, flying out in all directions. A wall outside the furnace wall contained a second knife-edge slit. Some of the atoms flying out of the furnace inevitably flew straight at this second slit, went through, and emerged on the other side to form a thin beam that was energized by light from the discharge lamp directed onto it by magnesium oxide reflectors. This was the beam with which Gould worked. The number of atoms it contained amounted to only a few thousand per second, and only 5 percent of those were excited to the metastable state. Nevertheless, it was a far higher excitation rate than anything he had tried before.

He sent the beam through two magnetic fields, one after the other. The first magnet bent the beam, sending it in different directions depending upon the state of the atom, then the second magnet bent it back again, focusing on a detector that consisted of a single white-hot wire connected to a galvanometer. The atoms falling on the wire would evaporate and leave electrons, which showed as a small current. Gould then had to apply to the beam before it reached the detector a radio frequency field. The right frequency would trigger the transitions between the hyperfine states in the atoms that he wanted to measure. But finding that frequency was hard. When he found it, the detector would show a drop-off in current, meaning the transitions had scattered the beam and not allowed it to refocus. It was a painstaking task, turning the frequency generator dial by fractions while watching every second for a drop in the detector current to be sure he hadn't simply passed over the correct frequency.

The thallium was gummy stuff, so after every few runs Gould would have to put on a laboratory smock and take the machine apart to clean it, scraping off the thallium and, when he was finished, replace the vacuum seals and evaporate magnesium oxide onto the lamp's reflectors to form a new shiny surface for bouncing the light onto the beam. This took two or three days, and then he would begin again.

He had the machine up and running one night toward the end of 1956 and was about to start twisting the radio frequency dial when Kusch walked into the laboratory.

"Oh, so you're ready to look," he said. "Let me help."

Kusch sat down and slowly twisted the radio dial while Gould watched the galvanometer attached to the beam sensor. Five minutes later, at about ten o'clock, the galvanometer's needle dipped—the beam had moved off the detector.

"Stop," he said. "Right there."

Kusch twisted the dial back a hair and the needle dipped again. At last, the transition was occurring. Gould checked the needle once again before he allowed himself to believe it. Then he jumped up and cheered and yelled. Kusch jumped up with him, his stentorian voice echoing out of the laboratory and down the halls so that students working late in other labs came to see what the excitement was about.

"Well, Gordon, finally," the professor said, and clasped his student affectionately about the shoulders.

They repeated the measurement several times that night. Over the next few weeks, Gould repeated the experiment from the ground up, recalibrating the oscillator that produced the frequency because the measurement was aimed at confirming a modification in the theory of the interaction of the atomic nucleus with the surrounding electrons. It had to be accurate to one part in a million, the equivalent of a clock that is accurate to within 30 seconds over an entire year.

With the figures he recorded, as 1956 drew to a close, Gould could at last look forward to completing his thesis and getting on with his life.

10

Even before his breakthrough, Gould had seen other possibilities in optical pumping. As early as the previous summer, when he was the only person in the country working with and trying to command the process, he had determined that it could be used to excite a maser. This would not increase its usefulness, but it would vastly expand the ways a maser could be made. Producing a population of excited molecules or atoms, rather than isolating ones already there, meant the maser would no longer be restricted to using an ammonia beam—the only material that could be used in the Townes-Gordon-Zeiger scheme.

This was a huge and significant step. It meant turning Boltzmann's distribution of atoms—the vast majority in the ground state, then ever fewer as they rise in the allowed stages of excitement—on its head. Gould's plan would create what is called a population inversion, a reverse in the usual order of things and a necessary first step to stimulated emission. Just as in the ammonia maser, the excited atoms could emit their radiation without having it all absorbed by those in the ground state, and the emitted photons would impinge on other excited atoms and stimulated emission would occur. The difference was, you could now employ virtually any element whose ground states contained microwave transitions.

At a molecular beams conference at the Brookhaven National Laboratory on Long Island in August 1956, he approached Charles Townes and described his idea.

Townes, with the maser, had begun to achieve international prominence in physics. The machine, despite its limited applications, had extended the boundaries of spectroscopic investigation. This, in Townes's view, was a more important step than high-energy physics and other subsets of the field that were considered more glamorous.

Townes's growing reputation owed much to his activities outside the laboratory. He had ended four years as chair of Columbia's physics department the year before. Previously, in addition to directing the Columbia Radiation Laboratory, he had chaired the visiting committee for physics at Brookhaven. He was also active in Washington, where he had chaired the Millimeter Wave Committee for the Office of Naval Research and later spent two years as a member of the Physics Advisory Committee to the Air Force. He tucked these administrative duties and committee memberships among far-flung academic posts, serving first as a national lecturer, in 1950–51, for Sigma Xi, the science fraternity, of which Gould also was a member. Townes had lectured at the University of Michigan in the summer of 1952, and at the Enrico Fermi International School of Physics in 1955, the same year he won a Guggenheim fellowship. He was a Fulbright lecturer at both the University of Paris and the University of Tokyo in 1955 and 1956, while on sabbatical. Townes's lengthening résumé and his list of scientific publications, contacts, and affiliations were the stuff of scientific celebrity. Had it not been for his vow to always save Sundays—he and his family attended services at Riverside Church, a popular multidenominational church with Baptist roots—as a sacred day when one didn't do "ordinary" things, it seems unlikely that he would have seen his wife and four daughters at all.

Townes by now had patented the maser, but only as an ammonia beam device. When Gould mentioned his idea for a light-excited maser, Townes reacted with what Gould thought was extraordinary interest. He asked Gould to lecture on the subject at one of the regular Friday seminars that brought physics students together at Columbia's Radiation Lab.

Gould made notes and delivered a lecture in December 1956, as he was perfecting the process of optical pumping in his thallium experiment.

Over the same fall months, when he was preparing his lecture, Gould peppered Townes with questions about patents. He remembered only too well his paralysis at conveying his ideas about the contact lens to Theo Obrig. Feeling certain that optical pumping would be a valuable addition to the maser, he wanted to be able to protect himself.

Townes advised him to record his plans in detail in a notebook and have it witnessed. This would establish when his idea was conceived.

Gould followed Townes's advice. During the year-end holidays, he wrote his ideas and calculations in a laboratory notebook and headed them, "Notes on the Principle and Some Applications of Differential Optical Depopulation of Atomic or Ionic Levels (Optical Pumping)." As soon as the holidays were over, he rushed into Townes's office to have the professor witness it. Townes signed Gould's notebook on January 3, 1957.

Gould also asked a postdoctoral fellow named Robert Novick to witness his notebook. Novick read it and signed on January 14.

■■

Gould as much as any beam of atoms seemed to have been energized by his experience with optical pumping. He was as productive as he had ever been, his head was full of new ideas, and he was confident in the soundness of his thesis information. In February 1957 he took and passed his Ph.D. exams.

Then, Kusch changed the rules.

The professor asked Gould to make another measurement. This time, he wanted Gould to determine the magnetic moment of the thallium atom in the metastable state. This would consist of two measurements—the magnetic moment of metastable thallium and, in the same magnetic field, the known moment of an atom such as potassium for comparison.

Gould was stunned at first, then furious. The measurement would add months to his thesis work, and he had already been at it for over five years, counting the time he had worked with Berman and Franken. He knew that Kusch only wanted him to do it because he had mastered optical pumping, and the opportunity to make the measurement would otherwise be lost. Gould understood the scientific impulse, but thought the request was unfair. At the same time, he owed Kusch for finding him the research assistantship that had sustained him after he lost his job at City College.

Ruth reacted more harshly than her husband. "He can't do that," she complained.

"He did it," Gould said.

"But you'll never get your Ph.D.," she railed. "We can't live on my salary forever."

"What am I supposed to do?" Gould said.

Disgruntled, he began the new experiment. The apprentice student working with him gave up in dismay, and left Columbia to pursue his doctorate elsewhere.

By the end of the summer of 1957, Gould had the new measurements. He was again ready to start work on his thesis dissertation, but now a new idea was beginning to distract him, and it was so compelling he could not deny it. What if, he thought, you could use optical pumping to amplify not microwaves, but light waves.

■■

Gould was not the only person to whom this had occurred. Talk among wave scientists in the halls of the Pupin Physics Laboratories, in other university and corporate research laboratories, and at the scientific meetings physicists attended, was about the prospects for moving maser action to visible wavelengths. It was an idea that was in the air.

The wavelengths of electromagnetic radiation vary widely. Imagine the undulations of sand dunes versus the fine teeth of a jeweler's saw. At radio frequencies, the distances from the peak of one wave to the next—the definition of wavelength—are relatively long, from the meter range to longer than a mile. They grow shorter as they approach and pass through the microwave region, where they measure in centimeters. They grow shorter still as they approach the infrared end of the visible spectrum, where they measure in microns, or millionths of a meter, and continue through the visible—starting with red and ending with violet—into the even shorter and invisible ultraviolet waves. Visible light waves are about 30,000 times shorter than microwaves.

Harnessing light waves presented very different problems from those that Townes had faced in amplifying microwaves.

Townes thought about these problems, and saw them as too difficult to solve directly. One would have to sneak up on them, so to speak, by learning to amplify waves in the interceding wavelengths first. Townes's initial thoughts, as he expressed them in a later deposition, "were to push down below the microwave range into the sub-millimeter wavelengths and then gradually beyond that further into the infrared. I felt that stepwise this development would be a natural one."

Nevertheless Townes, like Gould, was thinking by the early fall of 1957 of how to apply the maser's principles to light waves.

On September 14, 1957, Townes wrote the words "A Maser at optical frequencies" on the crosshatched graph paper of a laboratory notebook, and drew beneath it a box. The box was lopsided, like a sagging tool shed, but it wasn't meant to be that way; Townes, despite his many talents, was no artist. His drawing showed a hole in one side for an input light, and a hole in the top for an output light. He noted that he intended a "glass box silvered on inside or outside with 2 windows for input and output light."

In Townes's conception, the light would enter the silvered, six-sided cavity and as the atoms decayed from higher to lower energy states, emitting photons in the form of light that would simultaneously excite and stimulate emissions from other atoms in the box, the light would shoot all over the place, like light reflecting from a mirrored ball. The light would be sharply defined in color and frequency, but not spatially, that is, it would not be formed into the uniform straight beams associated with laser light today.

Townes described the boxlike device, similar to the outward appearance of the maser, as a maser at optical frequencies because the maser was the only wave amplification device then in use and he had invented it. But "maser at optical frequencies" was a grammatical kludge, a contradiction in terms, since the acronym refers specifically to microwaves, which are not visible and therefore do not exist at optical frequencies.

He added thoughts to the notebook over the next several days, dating each new entry. On September 16, he suggested the power needed in an input lamp. On the 19th, he jotted some entries about energy levels of excited thallium. He signed the notebook the same day, and had it witnessed by Joseph A. Giordmaine, who had been one of his graduate students.

A month later, Townes approached Gould with some questions, once calling him from the laboratory where he was at work. He wanted to know about the intensity levels Gould was getting from his thallium lamps. Townes recorded two contacts with Gould in his notebook, one on October 25, the other on October 28.

By then Gould's thinking had undergone a transformation. He had been only mildly interested at first, thinking like Townes that amplifying light waves was simply another step in spectroscopy. As Townes put it, "I was primarily interested in spectroscopy, and that would have been a very good spectroscopic tool. I didn't foresee a lot of energy at that point. I wanted only a few milliwatts, which was plenty to do the scientific work that I wanted to do."

Gould, however, realized that a great deal of energy could be obtained from stimulated light waves. He realized that the energy of photons rises in inverse proportion to the wave's length; that is, the shorter the wavelength, the greater the photon's energy. A microwave 10 centimeters, or about 4 inches, from peak to peak, when amplified produces photons coherent in time—that is, the waves are the same length and in step with each other. Otherwise, however, they are hardly earth-shattering. A visible light wave, far too small to measure with the naked eye, absorbs and emits photons that, Gould calculated with growing excitement, are a hundred thousand times more powerful than those of a microwave. This was power you could use. What was missing was a way to select a single spatial mode during the amplifying process, to single out waves that were not only marching in step but all moving in the same direction. Then you would have a beam that you could point, and you could focus, direct, and manipulate that power. The question, Gould realized, was how to isolate a single beam.

Townes's questions in late October awakened Gould to the extent of the professor's interest in amplifying optical wavelengths. Gould was helpful and provided the information that Townes wanted. But on the day of Townes's second contact, he went home that night and told Ruth, "I think he's thinking about the same thing I am. I'm going to have to get busy and write it down if I'm going to get a patent."

With that, he retreated to his study in the apartment on Townsend Avenue, where he pored over formulas, books, and papers until late into the night. For the next several days, Ruth found him preoccupied. When he wasn't deep in thought or scribbling calculations in his notebook, he acted like a child worried that a playground bully was about to snatch a toy. It annoyed Ruth that he wasn't working on his dissertation. He had been at his doctoral work far too long already. But Gould, hunched over his papers, seemed to think he was in some kind of race with the professor.

The insight, when it came, struck Gould with the force of revelation. It was a Saturday night. He sat up in bed, marveling at its perfection. Like all good solutions, it was simple and obvious, once you thought of it. It had been there all along, waiting to be seen for what it was.

He got up, leaving Ruth asleep, and went into the kitchen to make coffee. He polished his glasses on the shirttail of his pajama top, and looked at the clock. It was late. The early November night was mild, and a window was cracked to let the steam heat escape. Townsend Avenue was quiet, but from the Grand Concourse two blocks away and the Jerome Avenue el, two blocks in the other direction, came the honks and rumbles of New York City, never fully asleep even in the lull between midnight and dawn. Waiting for the pot to percolate, Gould lit a cigarette. The flame of the kitchen match wavered, and he noticed that his hands were trembling with excitement.

When the coffee was made, Gould poured a cup and went into his study. He sat down and thought it through to make sure he wasn't making a mistake. He wasn't. His mind raced as the dimensions of the breakthrough continued to reveal themselves. He could achieve, in an instant, temperatures hotter than the sun. He could cut holes in steel, cause chemical reactions, perhaps trigger nuclear reactions. He could communicate over enormous distances. After several minutes, he opened a lined laboratory notebook, the kind you can buy in any college bookstore, and started to write.

His fountain pen scratched on the cheap paper as he wrote, in a neat, right-slanting hand, "Some rough calculations on the feasibility of a LASER: Light Amplification by Stimulated Emission of Radiation."

Under that he drew a line, wrote some more, and sketched a diagram. It looked right to him, and he continued.

Gould worked through the night. The ashtray overflowed and the coffee pot was empty when Ruth came into the study wearing her flannel bathrobe. She leaned past the potted plants and raised the window blind. Gould blinked in the light.

"Are you working on your dissertation?" she asked with a yawn.

Gould lifted his glasses off his nose and rubbed his eyes, red-rimmed from smoke and lack of sleep. "I've got it," he said. "Look. This is the key, right here." He pushed the open laboratory notebook toward her and tapped his finger on the page.

She looked where he pointed, and read his handwriting: "Conceive a tube terminated by optically flat partially reflecting parallel mirrors." His drawing of two end pieces connected by lines seemed to be a schematic of a tube.

"The light will reflect between these mirrors," he said, his voice rough and weary, but excited. "It will form a plane wave. The light will be spatially coherent."

"What do you mean 'coherent'?"

"Light from a light bulb is incoherent," Gould said. "The light waves scatter all over the place. Spatially coherent light, where the peaks and troughs of the waves are all lined up and going in the same direction, would form a thin beam in a pure color."

"I see," Ruth said, but she didn't. Despite her Ph.D. in biophysics, she knew little about optics or quantum mechanics and had no conception of how molecules and atoms interact with radiation, including light. She withdrew to the kitchen and started making a fresh pot of coffee.

Around ten o'clock, her brother John came down from his apartment upstairs to see if they wanted to go out for breakfast. There was a place they all liked on Jerome Avenue in the shadow of the el. But Gordon shook his head. "You go ahead," he said. "I'm going to keep working."

He was still at work when they returned. John went back up the narrow stairs to his apartment. Ruth came into the study, carrying Sunday's *New York Times*. The rocket scientist Wernher von Braun was in the headlines, saying the United States was "well over five years" behind the Soviet Union in earth satellites. It would take a "well-planned, long-range space flight program" for America to overtake the Soviets, von Braun said.

The country had been reading similar headlines for a month, since the Soviets launched an earth satellite called Sputnik on October 4. Nothing had shocked the United States as much since the Japanese had attacked Pearl Harbor. Americans had taken their technological superiority for granted since the atomic bomb ended World War II. Now ordinary Americans rushed into their back yards at night and craned their necks for a sight of the satellite. Its beeping radio signals were played over and over on the news. Within days, as announcers on radio and television spoke with concern about the "technology gap," Congress had begun debating new funds for science education.

The shock had deepened just the week before, when Soviets launched Sputnik II, carrying a dog, into the heavens. Gould remembered it well. Just like that, complacency had changed to a national mood of desperation, and technological advancements of all kinds took on oracular significance.

"What are we going to do this afternoon?" Ruth asked. "It's a beautiful day." Sometimes they drove to Scarsdale on Sunday afternoons to play bridge with Gould's parents.

"I have to keep working," he said. "I have to get this down and get it dated. It will be important when I apply for a patent."

"A patent? What about your thesis?" she complained. "If you have to work, you at least could work on that."

Gould sighed. He lit a fresh cigarette, drew smoke into his lungs, and on the exhale said, "Believe me, this is more important."

"How can it possibly be more important than a doctorate?" she said. "How else are you going to get a teaching or a research job?"

"I've had jobs like that. I don't like them very much."

"Well, you have to make some money sometime. What do you think it will do, anyway, this so-called laser?"

"I think it will do a lot of things. The beam could go on for miles with little dissipation, I mean, without spreading out very much. Think if you could do that with a light bulb. It would be a hundred thousand times more intense. You could focus it to heat and cut things. Use it for long-distance signaling. You could measure distances by calculating with the speed of light. But I have to finish writing it up. That's the first thing I have to do before I can get a patent."

"Why don't you write it up for your thesis?" Ruth asked.

"Kusch wouldn't let me. It's too applied. He just cares about pure science. It would be beneath him to care about something like this, that could really do something."

"Well, I still think a Ph.D. is more important than this thing, no matter what it does," Ruth said. But she left him alone, and Gould kept working.

He worked through the weekend and into the next week, consulting reference books, filling large sheets of white paper with calculations, and writing in longhand in his notebook.

By Wednesday, November 13, Gould had written nine pages. The first six described how his "laser" would work, using potassium vapor as the medium. In the simplest possible terms, potassium atoms in a clear-sided tube would be optically pumped to a high energy level by the resonance light from an external potassium arc lamp. The atoms would give up their energy while coherently amplifying a light beam. This beam would bounce back and forth between the mirrors at each end of the tube, exciting other atoms as it did so. The energy would grow into a powerful beam of pure color and direction, and eventually break through the partially transparent mirror Gould had envisioned at one end.

Gould's calculations were rough, as he had written at the beginning, but he could picture the phenomenon. By seeing it, he had made a leap that so far had eluded scientists who relied on theory and mathematics.

The final pages described what Gould thought the laser could do. He wrote at the top of page 7, "Brief statement of properties and possible uses of the LASER."

He thought it would make a reasonably good spectrometer. Addressing the narrow beam, he wrote, "The (coherent) beam of light could emerge from the partially transmitting mirror as a wave which was plane to within a fraction of a wavelength. At a distance of one kilometer the beam would have broadened [approximately] one centimeter. Thus the beam could travel long distances essentially unweakened. Applications to communications, radar, etc. are obvious."

But "perhaps the most interesting and exciting possibility," he wrote, "lies in focusing the beam into a small volume." The power of an ordinary light bulb, 100 to 1,000 watts, "could be generated in a large volume and focused into a small volume with a tremendous factor of energy concentration" that would instantly heat solids or liquids placed at the focal point to temperatures as great as the core of the sun. He envisioned performing difficult chemical reactions, and "if the substance were heavy water, nuclear fusion temperatures could possibly be reached before the particles were dissipated."

Beyond that, he saw his "LASER" as an amplifier of light signals with applications to television and astronomy.

When he finally was finished, he wrote, "the foregoing 9 pages were written on or before Nov. 13, 1957." He set off up Townsend Avenue to the candy store where, he knew, there was a notary public.

The temperature was pushing sixty, and the sky was mostly fair. Gould walked bareheaded along the street of small four-family buildings interspersed with larger ones. Gray was beginning to show in his brown hair. He felt weary from his labors. At the same time, the smoke he inhaled from his ever-present cigarette was flavored with the taste of expectation.

Jack Gould at the candy store looked up as Gordon entered. The notary knew Gordon because he and Ruth often shopped at the store, but they weren't related.

"What can I do for you today?" he said.

"I've got something to be notarized." Gordon produced the laboratory notebook and opened it. "I wonder if you'd put your stamp on every page?" he asked.

It was an unusual request, and Jack Gould looked at the page before him. The drawing, the writing filled with numerals and symbols, the equations, and the footnote at the bottom of the page so filled the space that there was hardly room for the notary to place his seal. He leafed through the following pages and saw much the same. "I'll have to put it in the margin there," he said.

"That's fine," Gordon said.

Jack brought out two stamps. One was the standard "sworn to and subscribed before me" stamp where the date would go, and the other was his own notary's seal with his license number, 03-1521950. He inked them and pressed them down where he found space, sideways in the left-hand margin near the top of the first eight pages. Gordon signed them, and Jack Gould added the date and his own signature. Only on the last page was there room for the proper horizontal alignment of the stamp. The notary inked it, pressed it down, and wrote his final signature.

Gould paid for a carton of cigarettes along with the notary fee, and headed home with his smokes and his signed and notarized notebook. He knew it contained something important. As he walked back down Townsend Avenue, Gould thought that this was what he had dreamed about when he had dreamed of being an inventor.

12

Gould's professor fitted a cigarette into its holder and struck a match, held it to his cigarette, then to Gould's as Gould squirmed in the wooden chair across the desk. Kusch's face and the round eyes magnified behind his glasses showed deep disappointment. It was March 1958, and Gould had come to tell him that he was leaving the graduate school.

"But you're so damned close," Kusch said, his voice booming in the narrow office. "You can't leave ABD." All But Dissertation was the term for people in Gould's shoes, who had cleared every hurdle toward a doctorate but the final, and most important, one: the dissertation on the thesis project and its defense.

"I have to," Gould insisted. "I don't have any money. I'm tapped out. I just can't continue."

The facts were slightly different. Ruth would have been more than happy to stake her husband to the few remaining months it would have taken to complete his thesis work. But Gould was now committed to a different course. He was convinced beyond doubt of the vast possibilities represented by what he called the laser, and he was beside himself to obtain a patent. The problem was, his brilliant insight into light amplification was not matched by a grasp of patent matters. Once Gould had had the notebook notarized, he had no idea what to do with it.

He fretted through the weekend holidays, added a page or two to the notebook with refinements in his thinking, and finally summoned up his courage and consulted with a patent lawyer.

Robert R. Keegan worked at Darby and Darby, an established New York patent law firm with offices on Lexington Avenue. Its senior litigator was a friend of Gould's family in Scarsdale; Keegan was only an associate, but he also was the only member of the firm who knew what a maser was. A transplanted Oklahoman, he was a rangy, broad-shouldered man who still had a look of the prairie about him, with an upswept pompadour and eyes that turned down at the corners. His middle Southern accent was discernible, but barely, and he had an elliptical way of speaking that circled the point but eventually closed in. The two met at Keegan's office in January 1958.

Gould was typically close-mouthed about what he had invented. He mainly wanted to know if it would be covered under Columbia's patent agreement. Government and military contracts funded much of the research conducted at Columbia, as at most major research universities. The patent agreement, which all university researchers were required to sign, stipulated that any inventions that emerged from government-paid research would be licensed to government agencies and branches of the military free of charge. Gould described the laser in the vaguest terms, and told Keegan he had invented it outside the boundaries of his thesis work.

If that was the case, Keegan said, the patent agreement would not apply. But Gould left the meeting with the wrong impression about a vital aspect of the patent law. He believed that he would have to build a working laser—reduce his idea to practice, in the law's terminology—before he could obtain a patent.

Building a new machine is one way to get a patent, but it's a requirement in only one case. So many—and so often harebrained—have been the schemes to produce perpetual motion that the U.S. Patent and Trademark Office requires a working machine before it will issue a patent on perpetual motion. The more usual way to obtain a patent in all other cases is to provide a "constructive reduction to practice." That is, write a patent application that would allow a person "skilled in the art"—someone who knows well the area in which the invention falls—to build it.

Gould had made a fundamental mistake.

The first thing he tried to do was figure out how much it would cost to build his laser. It didn't take him long to figure out he couldn't do that and stay at Columbia. He would need a basic laboratory, including ovens, glass blowing equipment, and spectrometers, and at least $20,000 worth of experimental time. The Columbia physics laboratories were well-equipped enough, but Gould was convinced, as he'd told Ruth, that Kusch wouldn't let him switch thesis projects at this late date. He didn't want to do it as a thesis anyway. He'd been at Columbia for almost nine years working in pure science. He already saw the laser as so much more than that. It was more than proof of a scientific theory, more than a thesis project. It was a machine, a useful machine that would do wonders.

Gould also knew that if he worked at Columbia on a wave amplification device, even one so vastly different from the maser, Townes would be involved and would likely get the credit. Since he had to build it anyway, he wanted to build it on his own, free of the shadow of Townes and the maser.

This meant he needed an academic or commercial research laboratory that would get behind him, a place where, within the confines of a job description, he could work on his invention.

Both the Stevens Institute of Technology, in Hoboken, New Jersey, and Rensselaer Polytechnic Institute in Troy, New York, told Gould he would have to get outside funding and also teach a course load as he pursued his research.

Then, late in February, Gould heard of a private research laboratory in Manhattan that was hiring. It went by the initials TRG, which stood for Technical Research Group. He made some phone calls, and soon was headed downtown for an interview. The company was expanding, landing new research contracts, open to new projects it could sell. Gould didn't lay out his ideas on the laser, but he got the feeling he could persuade this eager and freewheeling company to give him the time to work on it. The interview went well.

Job offer in hand, he had gone to give Kusch the story that he was leaving school as a matter of necessity. He vaguely promised that he might come back to finish his dissertation and complete his Ph.D. requirements. But for the time being, he said, he had no choice but to leave.

■■

Lawrence Allen Goldmuntz was a young Ph.D. in electrical engineering when he started TRG in 1953. He was a bundle of barely contained energy, with wiry black hair and eyes that seemed frozen in an expression of permanent surprise. In the three years since receiving his doctorate at Yale—where he had earned bachelor's and master's degrees as well—he had worked as an engineer at the General Precision Laboratory and at Nuclear Development Associates, in White Plains, New York.

His work at Nuclear Development brought Goldmuntz in contact with Bernard Lippman, a physicist who had studied under Julian S. Schwinger, a Columbia Ph.D. whose work in quantum electrodynamics was destined to win a Nobel Prize. Goldmuntz was "a young and insecure bachelor," while Lippman was already an accomplished and prominent physicist, but they were united in their disenchantment with the company's management. They decided, in Goldmuntz's words, "If these clucks can run a company, so can we."

Before long, TRG was doing business out of three rooms on West Forty-fifth Street in Manhattan. Goldmuntz won a contract to do theoretical research on missile guidance for the Applied Physics Laboratory in Washington, and Lippman brought in a contract to study electromagnetic scattering. TRG was in business. But the partnership was short-lived. They argued, and Lippman resigned, leaving Goldmuntz with the company and a headful of anxiety about whether he would be able to keep it afloat.

Meanwhile, in Boston, a pair of Ph.D.s in math, from Harvard and MIT, respectively, were reaching the same conclusion about working for others that Goldmuntz had come to a year earlier. Jack Kotik and Alan F. Kay had met at the beginning of their doctoral studies, and had both gone on to work for Macmillan Laboratories, in Ipswich, Massachusetts. After working on radomes, the coverings for radar equipment mounted on the noses of airplanes and missiles, for upwards of two years, Kotik and Kay decided they wanted to start a company of their own.

Somebody at the Air Force's Cambridge Research Center knew Goldmuntz and suggested that they contact him. The three decided to join forces. Kotik moved to New York, and Kay remained in Boston to operate a branch there.

The business grew rapidly. In the postwar scientific explosion, government contracts grew on trees. Kotik and Kay brought work from the Cambridge Research Center. Goldmuntz still had the Applied Physics contract, and quickly added others. The company outgrew its midtown offices, and moved downtown to 17 Union Square West. Adding experimental work to its theoretical beginnings, the company took over more offices and converted them to laboratories. By 1956, TRG occupied two floors on Union Square, with laboratories scattered throughout the space. The company appropriated the building's roof for antenna measurements as well as touch football games. Kotik, a rangy man with bulging eyes and a love of outdoor sports, was the only company principal to join the games, which sometimes ended when the football sailed off the roof and landed in the street.

13

Before Gould abandoned his thesis and started work at TRG, Charles Townes had brought his brother-in-law Arthur Schawlow into his attempt to figure out the secret of amplifying light waves. Townes was stuck on his box. He still thought it was not all that important to get the machine to choose a single mode; that is, isolate a single beam. Nevertheless, he recognized the cavity— the shape that would contain and reflect the waves—as a problem, and in a roundabout way had sought Schawlow's help.

Schawlow in 1951 had moved from Columbia to Bell Labs after Townes's elevation to department head had brought Columbia's anti-nepotism policy into play. He was working out of his field, in a group originally headed by the transistor's co-inventor John Bardeen, that was conducting low temperature superconductivity experiments. After six years, Schawlow was jaded, and the physicist under whom he worked had turned to Townes for help.

Albert M. Clogston told Townes, "Art is having a hard time right now. His work isn't going too well. Why don't you have a talk with him?"

Townes had maintained his relationships at Bell Labs. He consulted with the scientists there and received a monthly stipend which required no accounting. He visited Bell Labs in Murray Hill, New Jersey, every week or two under this arrangement, and on his next visit, in the fall of 1957, he sought out Schawlow and took him aside for a talk. Townes's approach was to act as something of a scientific cheerleader. Searching for something that would spark Schawlow's interest, he told him he had been thinking about the possibilities for amplifying light waves.

Schawlow reacted enthusiastically. He told Townes he had been thinking about the same thing.

Out of their mutual interest grew an informal pact to work together.

Through the end of 1957 and into 1958, Townes continued his regular visits to the Murray Hill laboratories, and after he saw the people with whom he was consulting on masers, he would spend half an hour with Schawlow. They brainstormed and kicked ideas around, without trying to do experiments. They settled on potassium vapor as a likely amplifying medium. But after that, the collaboration stalled because they were stuck with Townes's box as a cavity configuration.

"We were thinking in terms of a box resonator like he had," Schawlow said. But whereas the maser Townes, Gordon, and Zeiger had built fit the dimensions of a standing microwave, this box had to be different. "You could hardly make it as small as the wavelength," Schawlow continued. "If you made it that small, there wouldn't be room to put any atoms in it. But we assumed it had to be of centimeter dimensions."

That left them with the problem Townes had anticipated when he had sketched the rectangular box in his laboratory notebook. "There would be millions and millions of modes in such a box," Schawlow said. "There would be an awful lot of them, and Charlie thought, Well, maybe it would pick up some mode that had the lowest loss [from reflection or refraction, the principal causes of light loss], or maybe jump from mode to mode. The light that was being emitted by this box would come out in different directions, would jump from one direction to another. But it would still be different from an ordinary light source."

Schawlow was saying that the atoms rocketing around inside the box would be doing so at random, exciting and amplifying others that would send off beams at all angles, each for a fraction of a second. What they had so far was the world's most revolutionary light show, but no means of bringing it under control.

Then, in February of 1958, Schawlow had the revelation that had visited Gould three months earlier. He was shaving one morning, and started thinking about mirrors. Suddenly his memory jerked back to a piece of equipment he had used in his thesis experiment at the University of Toronto. The Fabry-Perot interferometer was a diagnostic tool used in spectroscopy to distinguish very small differences in wave length. It was essentially a tube joining two opposing mirrors.

Townes arrived at Bell Labs the next day. He made his rounds and then sat down with Schawlow for their regular meeting. Schawlow told him, with muted excitement, that he thought he had solved the problem that had been stumping them as long as they conceived of the "optical maser" as a box. He showed Townes a sketch and said, "I think this is a good way to select a mode."

Townes looked at what Schawlow had drawn, and wondered immediately why he hadn't thought of it himself. Two simple mirrors, he thought. It clarified things greatly. "You're right," he said. "It's better than you think. It's much better because the light will be amplified. It will bounce back and forth a number of times, so it has to be really directional. And it won't amplify any beams that go off to the side."

Before their session ended, Schawlow and Townes reached the next realization, which was that one of the mirrors had to let some light through while the other had to do nothing but reflect.

They had come up with precisely the same configuration as Gould had sketched and described the previous November—"a tube terminated by optically flat partially reflecting parallel mirrors."

■■

Townes, playing the part of the good brother-in-law and still concerned that Schawlow was, as he put it, "maybe a little bit at loose ends," now suggested that they collaborate on a paper describing what Townes called an optical maser.

Townes's thinking had changed since the original maser. In those days, eager to test the technological possibilities that emerged from World War II, many physicists were writing about what they were going to do. New ideas were finding their way into print in the pages of *Physical Review* far more often than those ideas were realized. The noted publication of the American Physical Society was being called *"Physical Preview"* by cynics, and Townes felt his idea had to be demonstrated before anything was published. This let a pair of Soviet scientists, Nikolai Basov and Aleksandr Prokhorov, into the picture, for by the time Townes, Gordon, and Zeiger had built their maser and published their work on it, Basov and Prokhorov had published similar, though less complete, ideas. Their publication stole a share of Townes's glory. This time, he was determined to be first.

He and Schawlow started work on their article, which they intended would form the basis of a patent application.

Townes had already patented the maser. That patent he had assigned to the Research Corporation, a nonprofit patent licensing firm that distributes its income among universities to fund research. It was a basic patent, one covering a fundamental process, and Townes had taken pains to ensure that, while it did not specifically mention light, it covered all frequencies. The original application had not described a means of exciting a medium, but Townes had taken care of that in January 1958 by filing a continuation application. That application added optical pumping to the maser's capabilities, in a manner identical to that described by Gould in the notebook he had given Townes to witness twelve months earlier.

Gould, caught up in the prospects of the laser, had never bothered to file an application on optical pumping of the maser.

Townes believed his work with Schawlow would simply extend the maser specifically into visible frequencies. But since Schawlow's role was crucial, and since Schawlow was working for Bell Labs, Townes decided any patenting of the new device should be handled by Bell Labs. Here he caught a whiff of trouble, which he wanted to avoid. He feared that the Research Corporation's patent lawyers, with whom he was in regular contact, might see things differently. So when he talked to them, he kept his work with Schawlow to himself.

When the brothers-in-law were finished with their paper, they ran it by Bell Labs' patent department. This was standard procedure. Papers submitted for publication always went to the patent lawyers first, so they could be screened for patentable devices. The paper wound up in the hands of Lucian Canepa, one of a dozen or so lawyers in the Physical Research division that handled semiconductors and optical developments. The division had prosecuted Bell Labs' groundbreaking transistor patent filing. The publications output of Bell Labs physicists was such that Canepa sometimes received several papers a day.

Canepa called Schawlow to discuss what he and Townes had written. He wanted to meet quickly, because the *Physical Review* deadline was imminent. Schawlow came to his office.

Canepa asked Schawlow how the so-called "optical maser" differed from a maser.

"Well, it's a different wavelength," Schawlow said. "It gets down into the visible range."

Canepa asked about its commercial possibilities.

Schawlow couldn't think of any. He scratched his head and said, "It's kind of theoretical, and kind of experimental. Charlie's coming over tomorrow, and he has a better idea about these kinds of things, so why don't I bring him in?' "

Schawlow took what he thought was the flavor of the meeting back to Townes. "They don't want to patent it," he said.

Townes was surprised. "Why not?" he asked.

"They said, 'Well, it's just a maser. It's a different wavelength, all right, but it's just a different wavelength.' They don't see that light has much to do with communications, and it wouldn't be much use to them," Schawlow reported. "They said, 'You can patent it yourselves, if you like.' "

Townes couldn't believe what he was hearing. "They are clearly wrong," he said. But rather than take up the suggestion that they patent it themselves, he added, "I don't think we should take advantage of the fact that they just don't understand. It would be unfair. We had better talk to them."

Within a day or two, Schawlow and Townes appeared together for a new round with Canepa. They were joined by Arthur J. Torsiglieri, who was assistant patent director for the Physical Research division. Townes gave a more communications-oriented flavor about what they were presenting. Although somebody wondered if clouds and rain would interfere with optical maser beams, enough was said that the patent lawyers decided they would file an application after all.

Torsiglieri reminded Canepa of *Physical Review*'s deadline. "You don't have very long," he said.

Canepa said he'd do his best, and drafted the patent application in a week. He sent the draft to Schawlow, who returned a few days later after apparently reviewing it with Townes.

"It looks fine," he said.

The application that Canepa filed with the Patent and Trademark Office on July 30, 1958, bore the title "Masers and Maser Communications System."

Schawlow and Townes then put the finishing touches on the article. A draft had already been circulated as a memorandum at Bell Labs. With the patent application submitted, they felt free to circulate preprints of the article at about the same time they submitted it to *Physical Review*, where the paper was received on August 26.

14

Gould also was furiously at work.

Since March 27, when he started at TRG, he had spent every spare moment searching background material for insights that he hoped would help him build a laser. He had scoured the scientific literature for everything he could find about the optical resonance lines of excited matter, looking for alternatives to potassium and the other alkali elements he had proposed as laser media in the November notebook. He was not altogether sure the intensities he had assumed could be achieved, and the alkalis presented experimental difficulties, attacking in their vaporous form the very equipment in which they were contained. Spectroscopy had produced much information, and Gould read until his eyes watered looking for corresponding spectral energy lines that would allow him to create a sufficient population of excited atoms that would radiate at optical frequencies when they were stimulated to emit their energy. His desk at home was piled with tomes. The titles he explored included *Resonance Radiation and Excited Atoms*, *Fluoresence and Phosphorescence*, *Collisions of the Second Kind*, *Diatomic Molecules*, *Physical Optics*, *Atomic Energy States*, *Theory of Atomic Spectra*, *Applied Optics*, *Atomic Spectra*, and *Optical Properties of Thin Films*. He pored over articles in TRG's library during his lunch hour, while Kotik and the technicians were playing touch football on the roof. He went through the wavelength tables published by MIT. He found approximately one hundred strong spectral lines. Then, he searched for coincidences between those lines and elements that might be optically pumped.

He began to build a second notebook, more detailed and extensive than the first. His aim was to resolve the experimental problems he foresaw, and make a laser easier to build.

Gould had to moonlight at this task. The job he'd been hired to do at TRG was something different. He was supposed to be working on atomic beam frequency standards. His boss was Richard T. Daly, one of TRG's senior scientists.

Daly had a Ph.D. in nuclear physics from MIT. He was a big man, hearty and bluff, all but bald, with protruding ears, and an overbearing personality that didn't bode well for his relationship with Gould. At his previous job, he had invented the cesium beam atomic frequency standard, a highly accurate atomic clock using oscillations of cesium atoms. Frequency standards had military applications because their extreme accuracy made them potentially valuable in air and sea navigation and in missile guidance. But they were still in their early stages of development. They remained large, kludgy machines in which the kinks had not been ironed out. TRG had a contract from the Army Signal Corps to work on improved standards that were smaller, more stable, and easier to use.

Gould knew the field. The optically pumped maser he had revealed in the January 1957 notebook Townes had witnessed was in fact a frequency standard using rubidium.

Working during the day on his assigned tasks, Gould turned to his laser calculations at night and on weekends. But the enthusiasm with which he greeted the results of his research at home could not be contained when he got to work. At TRG, he worked with a theoretical physicist named Maurice Newstein. Newstein, one of TRG's earliest employees, like Daly had earned his doctorate at MIT.

Newstein's role in the laboratory with Gould was to observe, and to take notes that would confirm or disprove his mathematical expectations. There were long periods of preparation, and during those times the two men got to know each other. Newstein was not financially ambitious; he had enough money for his needs and was happy with his work. One day, Gould startled him by saying that his goal was to invent things and become a millionaire.

In May 1958, Newstein and Gould were experimenting on a particular kind of rubidium frequency standard. Seeking a microwave resonance that had to be photoelectrically detected, they worked in a darkened laboratory. They were working and talking in the dark one day that month when Gould told Newstein he was working on something very exciting, something extraordinary.

At first, he hedged and said he couldn't say anything more. Then, in a disembodied voice that Newstein remembered more vividly because he heard it

in the dark, Gould gradually revealed that he was working on an optical extension of the maser. "It's revolutionary," he told Newstein in his excitement.

Newstein agreed that it certainly would be revolutionary, if it could be
done. He knew, as virtually every physicist did, that people were talking about
it, that it was in the air.

Gould kept talking, and Newstein realized he wasn't hearing idle ram-
blings, but that Gould had thought it through in great detail. Newstein had
never heard anybody discuss nearly as many possibilities for achieving laser
action as Gould was talking about there in the dark. And the applications
Gould mentioned were really quite amazing, applications that could be
achieved by focusing to intensities that were many, many orders of magni-
tude higher than anything that had been achieved before, and applications in
long-range communication, especially for outer space. In all the talk about
pushing maser action into the visible, these were things Newstein had never
heard mentioned.

Gould continued to talk. He described optical pumping as one method of
achieving a population inversion, the necessary swarms of excited atoms that
stimulated emission requires, and then described another. You could do it with
collisions of the second kind, he said. These were collisions between atoms
(collisions of the first kind being collisions of electrons with atoms), in which
an atom in a lower, relatively long-lived, state of excitement called a metastable
state transfers its energy to a second atom. If you put the right mixture of gases
between a pair of electrodes and ran a current between them to create a discharge in the gas, the collisions between the two kinds of atoms would give you
enough excited atoms in one of the gases to begin the chain of stimulated
emission. He didn't have the gas mixture quite worked out yet—krypton was
one, and maybe xenon, or zinc—but he knew it could be done.

Newstein had heard none of these ideas, and here was Gould describing a
whole list of them.

Newstein wasn't the only person with whom Gould shared his thoughts
about the laser. Warm weather took him and Ruth to his brother Geoffrey's
home near the Connecticut shore to ready the boat for sailing season. Describing the laser to Geoffrey and his wife, June, Gould had to shine a flash-
light on their garage door to try to give them an idea. Alan Berman, Gould's
former laboratory partner, had a better time understanding what Gould was
working on when he told Berman about it at a Fire Island beach party.

In June, Gould took a month off from TRG, saying he wanted to work on his
thesis. Instead, over Ruth's objections, he worked on the laser.

Late in August, with the Labor Day weekend approaching, Gould showed the first twenty-three pages of his refined notebook to Lawrence A. Wills, a physics professor at City College with whom Gould had remained friends since his teaching days. Wills signed that he had read and understood them.

∎∎

The more his ideas advanced, the more Gould thought about the patent or patents he hoped to obtain. He was wondering how to separate what he had invented on his own from what would evolve from his job at TRG.

Larry Goldmuntz had held the title of president since the research group changed from a partnership to a corporation the year before. He was learning gradually what Gould was up to. The first inkling had come during the spring, when the company's personnel officer told him that Gould would not sign the standard patent agreement. It was boilerplate, much like Columbia's, providing that inventions developed under government contracts would be licensed to government agencies at no cost.

TRG remained small enough that Goldmuntz knew most of the employees, and he knew Gould, although not well. He had no experience with employees not signing the agreement.

"He's got to sign it," Goldmuntz said.

"He won't."

"He's got to, no ifs, ands, or buts."

The personnel director shrugged and looked helpless.

Goldmuntz knew Newstein had been working with Gould, and he stopped by the theoretician's office for a chat. "Do you know what this is all about?" he asked.

Newstein, a contemplative man given to long silences, told Goldmuntz that Gould had invented something before he came to TRG, and wanted to preserve the rights to his invention.

"Fine," Goldmuntz said. "Problem solved. Tell him to write down what he invented, and he can keep it. Have him write it out, and we'll agree that it was invented prior to his coming here." He didn't know what it was, and he left Newstein's office thinking that Gould was being awfully troublesome about something that was probably the technological equivalent of the buggy whip.

Newstein took his conversation with Goldmuntz back to Gould, and Gould signed an interim patent agreement that excluded inventions he had conceived prior to joining TRG. But he refused to tell the company what he was working on.

Gould's reticence pricked Goldmuntz's curiosity. He continued to ask Newstein about Gould's invention. Newstein reported that Gould had not yet finished his description of it. "He says he's still writing it up," he told Goldmuntz.

"Just a damn minute," Goldmuntz finally exploded. "How long is this going to go on? The company is liable. He's working on a government contract, and if he invents something we have to be able to convey it. We have a responsibility to convey it to the government, and if we can't, they can go after us. This can't go on forever. What is it, anyway?"

"He says he can push the maser up into the optical range," Newstein said. "He calls it a laser."

"Do you know what he's talking about?"

"Yes, but he'd be better at explaining it."

Gould didn't want to talk to Goldmuntz. Liberal suspicions about corporate power still lurked in his psyche, and he worried that his invention might be stolen. But Newstein prevailed on him, and the three met in Goldmuntz's office. By now it was September. The interim patent agreement was expiring, and Goldmuntz at least wanted it renewed. He said to Gould, "You've got to sign this damn agreement. What the hell is it you're working on?"

"I have a way to generate coherent light," Gould said.

Goldmuntz leaned forward, his bushy eyebrows arched. "You've got my attention," he said.

Gould explained the laser, its power, and the range of applications he foresaw.

Goldmuntz was skeptical. He wasn't a physicist, but his background in electrical engineering gave him some insight. He knew from having worked on a rangefinder for the Signal Corps' Frankfort Arsenal what a directional beam and a narrow spectrum meant, and he knew that coherent waveforms—in radio, at least—could carry a lot of signal information and made for good antenna reception. As he listened to Gould explain his scheme for this mysterious laser, he experienced no Archimedean "Eureka!" moment that allowed him to see the concept full-blown in his own imagination. But Gould's presentation was persuasive. Goldmuntz thought that the ideas hung together. Seeking further guidance, he brought Kotik and several other physicists at TRG into a session with Gould. He was prepared to hear them tell him Gould was crazy.

Gould stood at a blackboard in a conference room and for more than two hours outlined his ideas and described the uses he foresaw.

The physicists, used to drier presentations, thought Gould talked with an unusual amount of drama and enthusiasm. Kotik, for one, had no idea what to envision when Gould was finished. He didn't know whether this laser would be an inch long, or thirty feet. But they were intrigued, and curious, and as they

talked among themselves after he left the room, they decided that what he had
talked about seemed reasonable.

Goldmuntz gave Gould the go-ahead to work on his laser ideas on company
time.

TRG's president quickly became Gould's enthusiastic ally. He sensed the
laser's importance in potential military contracts, and soon angered Dick Daly
by pulling Gould off his frequency standard work and ordering him to devote
full time to the laser, a full-court press with the company's backing. And he resisted distracting him with talk of a patent agreement. "Don't worry about it,"
he said. "We'll talk about that later. Right now we've got to get cracking."

∎∎

TRG was growing, but it remained too small to underwrite the experiments it
would take to build Gould's machine. Goldmuntz asked him to prepare a proposal to submit to government agencies and private corporations. He didn't
know about the work that Schawlow and Townes were doing at Bell Labs, but
he assumed that with the laser a hot topic in scientific circles, TRG would inevitably be in a race with major research laboratories, and the only way to compete was with outside funding.

Gould set to work, and at the same time continued his research to put the
finishing touches on his second notebook.

Gould worked hard, but Goldmuntz found it difficult to get him to entirely
forsake his recreations. His summers since he was a teenager had focused on
sailing, and the call of his boat on the Connecticut shore was hard to resist. He
and Ruth had become friendly with Newstein and his wife, an anesthesiologist,
and they went sailing together. Gould also made time to go to the occasional
beach party. Goldmuntz had to push him to finish his notebook and write the
proposal. From Gould's perspective, Goldmuntz wanted him to give up the
material before it was complete.

Sailing season was over in the fall when a fire marshal visited TRG's space
on Union Square. It was a routine visit, but it took on new dimensions quickly
when the marshal learned that the company made its own vacuum tubes, a
process that required hydrogen braising. The combination of hydrogen and
oxygen produced a high-temperature flame with which workers melted an al-
loy sealing glass to metal in the vacuum tubes. The marshal told Goldmuntz he
knew a fire hazard when he saw one. And questioning Goldmuntz further, he
learned that a hydrogen leak can become explosive.

TRG's insurance company also was getting itchy at the work its client was doing in densely populated Union Square. Soon afterward, the lab made plans to move outside the city.

The company settled on Syosset, a town about thirty miles from Manhattan near the north shore of Long Island, where it took a 15,000 square foot building in a new industrial park with cookie cutter streets and scrawny new shrubs. Appropriate to its high-tech mission, the address was 2 Aerial Way.

TRG's scientists, technicians, and other employees were still unpacking as Gould put the finishing touches on his notebook. It was a document of eighty-seven pages, which he showed to Daly on December 1. Daly was still miffed about Gould having been snatched from his control, and the importance with which the company viewed his idea. Nevertheless, he wrote "read and understood," and signed it on the back page and on several inserted pages where Gould had made corrections.

Gould signed it the next day to signify that it was finished. Two weeks later, on December 16, the companion proposal was typed and in the mail. As a courtesy, Goldmuntz sent it to the Aerojet-General Corporation in Sacramento, California, a company that owned 18 percent of TRG. Copies of substantially the same 120-page proposal, seeking $300,000 in research funds to study lasers and laser properties, went to the Air Force Office of Scientific Research and the Army Signal Corps.

■■

Townes's and Schawlow's article appeared in *Physical Review* at almost the same time, on December 15, 1958. "Infrared and Optical Masers" discussed both multimode devices such as the box Townes had originally envisioned, and the single mode generator that Schawlow had come up with. It discussed both potassium and cesium vapor as amplifying media. It was a purely scientific article that mentioned none of the imaginative uses that Gould had foreseen for the device.

Nevertheless, it sent waves crashing through the scientific world. To the establishment that had never heard of Gordon Gould, Townes had created the maser and now he had taken the next step. Amplifying light was no longer only "in the air," but on paper, and at research facilities throughout the country physicists headed to their laboratories to see if they could convert theory to practice.

15

Gould and Goldmuntz welcomed 1959 by embarking on a round of meetings aimed at selling Gould's proposal. Things did not go well at first.

At Wright-Patterson Air Force Base outside Dayton, Ohio, home of the Air Force's scientific research arm, a physicist insisted that what Gould was proposing was impossible. At Fort Monmouth, New Jersey, Gould gave a detailed and enthusiastic presentation to officers and scientists of the Army Signal Corps, and Goldmuntz saw as he watched them that they simply did not believe the laser could do what Gould said it could. Back at TRG, they learned that Aerojet-General also had passed on the proposal.

Goldmuntz's ace in the hole was Richard D. Holbrooke, a physicist from California's Rand Corporation who had been assigned to the Defense Department's Advanced Research Projects Agency. Then in its infancy, ARPA was staffed partly by scientists assigned by the major defense contractors like Rand, Hughes, and Westinghouse. Goldmuntz had floated other proposals past Holbrooke, and even when they had been unsuccessful he had found him receptive. He called Holbrooke with a brief description of what TRG wanted to do.

"Come on down," the physicist said.

Goldmuntz sent the proposal ahead. Then he and Gould flew to Washington and took a cab across the Potomac to Virginia and the Pentagon, where Holbrooke met them.

"I have a confession," Holbrooke said after they had exchanged greetings in his office. "Your proposal looks interesting, but I don't know enough about the process to be sure. I've got somebody else I want to bring in. His name's Paul Adams, he's a patent lawyer who's on staff with us."

Gould, always protective, wanted to know if Adams would keep the proposal to himself.

"Oh, yes. You can be sure of that."

Gould and Goldmuntz assented, and Holbrooke said, "Okay. But before I ask him to come down, let me tell you about him, because I don't want you to be offended." He told them that Adams was big, loud, and often crude. Then he went to get him.

A few minutes later they heard big feet clattering down the hall. A booming voice said, "These guys know how to make coherent light? They'll have all the secretaries in this office menstruating in phase."

A big form filled the door, and Adams entered the room, followed by Holbrooke. "You guys know how to generate coherent light?" Adams demanded.

"Well, yes, we think so," Goldmuntz said.

"Okay, you've got an hour."

Gould made most of the presentation. Drawing from his proposal, he talked about light beams that could make accurate measurements over distances one hundred times longer than before, beams that could be used as infrared searchlights, giving a surface a mile from the source the same illumination as a one-hundred-watt light bulb at five feet. With more power, he said, such a beam generated on the moon would be readily visible to the naked eye on Earth, and the intensity of a laser beam arriving at Earth after the 50-million-mile journey from Mars would still be great enough to be readily detectable. All of this, he said, proposed uses in long-distance communications. He also suggested lasers could be used in TV projection, active radar systems, and long-distance alarm systems applied as burglar alarms, smoke detectors, weather detectors, traffic and railroad control systems, and in manufacturing process control. Turning to the intensity of a focused beam, Gould offered calculations showing vastly greater intensities than those available from a carbon arc lamp, and said that meant lasers could be used in high-temperature, high-speed chemical processing, X-ray technology, and possibly—as he had written in his first notebook—triggering nuclear reactions.

When the hour was up, Gould wasn't finished. "Missile guidance systems," he said. "Maybe even knocking missiles out of the sky." By then he had gone beyond his proposal and was talking off the top of his head.

"I think you bastards know what you're talking about," Adams said finally. "I think we should give you a contract, but I've got to clear this through the Killian Committee."

He referred to an oversight committee usually mentioned by the name of its head, Massachusetts Institute of Technology chemist and president James Killian. Adams told them he'd have an answer in three weeks.

Goldmuntz returned to Washington alone at the appointed time in March 1959. Adams reported that the Killian Committee had given its blessing. He read to Goldmuntz a letter lauding the proposal for its many good ideas, including direct laser use in rangefinding, communications, and target designation, and the possibilities suggested by its ability to bring enormous heat to bear instantly. Gould also had suggested medical uses, but Goldmuntz had talked him into setting those aside from the proposal, and saving them for his patent application on which the company also was hard at work.

Adams then read the letter's final paragraph, which suggested ARPA give TRG the $300,000 contract it had sought.

Goldmuntz's brain was exploding with dollar signs. Three hundred thousand was a lot of money for TRG. But a wave of dismay swept over him as he heard Adams growl, "I didn't ask them how much money we should spend. I asked them if the ideas were good." He watched in shock as the blustery patent lawyer leaned over a table, placed a ruler across the bottom of the page, and ripped off the final paragraph. Then Adams taped what was left of the letter to another sheet of paper, and substituted his own handwritten recommendation: "I, Paul Adams, think we should put $1 million into this program."

Straightening, he told Goldmuntz, "Give me ten days to get confirmation on this."

Goldmuntz flew back to Long Island in a state of extreme excitement, arriving in time for his one-year-old daughter's first birthday party.

■■

Confirmation arrived at the end of March, in the form of a project acceptance letter to Goldmuntz from ARPA. The letter was unusual, in that it was full of compliments. The agency's L. P. Gise had written, "Your proposal displays keen ingenuity and it promises a vital contribution to this nation's defense effort."

Not everyone at TRG was thrilled at the company's approaching windfall. Dick Daly, Gould's nominal boss, had seen the laser proposal as pie in the sky, and had even considered asking the company's board of directors not to submit it.

Now that a contract with ARPA was all but a reality, Daly took the highly unusual step of asking the board to turn it down. Goldmuntz believed that

Daly felt threatened by the attention the company was giving to the laser. He didn't like Gould and resented his success, but what he said was that he wasn't convinced that TRG could do what the contract called for.

Goldmuntz said, "I don't know, either. All I know is, they're going to give us a million-dollar check, and I say, 'Let's go.' "

Rather than back away from Gould's proposal, the board voted to reward him with company stock options.

Anticipation produced a frenzy of activity at TRG. The company now knew it would be racing against America's most formidable research giants, the likes of Bell Labs, Hughes Research, and Westinghouse, to produce a working laser. The million-dollar contract—actually $999,008—assumed that TRG would investigate the various laser approaches Gould had outlined simultaneously rather than sequentially. The company set about equipping laboratories and hiring new people.

■■

Charles Townes was one of the people who had been impressed with TRG's proposal. He had seen it early on in his capacity as a member of the Air Force's Scientific Advisory Board. In fact, TRG had specifically asked for his review after Gould and Goldmuntz encountered resistance in their pitch to the Air Force office of Scientific Research. Townes, the company knew, would understand the science. He had recommended the proposal, and ARPA was funding the contract through the Air Force's scientific research arm over the objections of the physicist in Dayton.

Townes seemed unable to believe, however, that Gould had come up with the ideas in the proposal on his own. He now remembered Gould not as the graduate student whose notebook describing an optically pumped maser he had signed, but as Kusch's somewhat indolent student who had left without finishing his thesis. There were necessary similarities between the proposal and the content of Townes's paper with Schawlow. Townes decided Gould must have seen one of the preprints that were floating around.

"He used a lot of ideas from our paper without giving any acknowledgement of it," Townes would say.

Townes still had no idea that Gould had written a notebook in November 1957 that had proposed laser excitation schemes and uses. He certainly knew, however, that Gould had grasped optical pumping. But as time went on, Townes would contend that Gould never did anything unless he, Townes, had done it first. He would tell oral historians at the University of California's Bancroft Library, "Gordon appears not to have started doing anything until he learned from our preprint that I was taking it seriously, and then he wrote down everything he could think of." Regarding Gould's initial notebook, Townes said, "I think that every time Gordon learned that I was doing something, he would dash to try to get it covered in his notebook. But then he never did anything if I wasn't doing something."

But he had. Gould's proposal to use collisions of the second kind—the collisions between atoms in a gas discharge in which energy is transferred—to excite a laser was unique. Townes hadn't thought of it, as he grudgingly conceded much later, after he and Gould came into conflict. "This I regarded as perhaps the most important and only suggestion that I felt was very novel or at least had not been openly discussed up until then," he said.

Coincidentally, one of Townes's former graduate students had also begun looking into the use of such collisions. Ali Javan confided to Townes, his mentor and friend, sometime around the end of 1958 that he was exploring the possibilities of a gas discharge laser. This, according to Townes, occurred around the time he received Gould's TRG proposal. Townes was struck by the same idea coming from two such different sources, and says scientific ethics kept him from mentioning to either Gould or Javan the other's work.

16

As TRG waited for the formal contract to be issued, Gould turned his attention to his patent application. He still had not signed a patent agreement, but he had come to trust Goldmuntz. They had talked off and on about the broad outlines of a pact that would divide the patent royalites, and while they agreed on its general structure neither man had had time to hammer out the final details.

Robert Keegan of Darby and Darby had been sketching the outline of the application since early December, when Gould finished his notebook, but a trial in Utica had diverted him at the same time as Gould and Goldmuntz were working at selling the proposal. Now all three of them could concentrate on the application. By the end of March, it stretched to over one hundred and fifty pages of descriptions, claims, and drawings.

Ominously, rumors reached TRG from Washington that the contract might be classified. When Gould heard them, he sought out Goldmuntz. "I hope it's not classified," he said.

"I know. It's a pain to work on those things," Goldmuntz responded.

"This would be more than a pain." Gould briefed Goldmuntz on his history, focusing on Glen, Prenski's study group, and his dismissals from the Manhattan Project and, later, from City College.

"We'll try to keep it from being classified," Goldmuntz promised. "But if it is, don't worry. We'll get you a security clearance. They wouldn't keep you from working on your own project."

Gould was reassured.

■■

ARPA's Holbrooke had sent the TRG proposal to a colleague at Rand, William H. Culver, whose role it was to be aware of new technology that might have military applications. He traveled around to research labs, talked to scientists, and brought home to his employer a notion of how the technologies he had discussed could be used in future weapons systems. Like Gould, he had seen military possibilities in lasers.

Culver read the proposal with mounting excitement. On March 30, 1958, he flew east from Santa Monica for a series of informal meetings about lasers. He carried a notebook in which he had written down eight questions about laser innovations. He had dated the pages because he thought his ideas could lead to new inventions. He was concerned, for example, that the amount of energy required to modulate a light beam was very high, and consequently inconvenient. Culver's idea was to use a Kerr cell, which changes the polarization of light and therefore could be used to turn oscillation on and off, as a way of modulating light without putting very much power on it.

Culver arrived in Washington as a late winter snow was falling. He met with Holbrooke, and then flew on to New York to meet with Gould.

The two hit it off right away. They both had been thinking along similar lines. As they talked Culver ticked off the questions he had written down, and explained his idea to use a Kerr cell to modulate the beam.

Gould replied, "That's a good idea. But if you put on a lot of voltage, then you could store up energy in this thing, and you could release it all at once in a giant pulse." He opened his hands like a magician conjuring a puff of smoke.

Culver recognized that Gould had taken his idea a step further. What Gould had explained would later be called the Q-switch, a reference to the quality factor of a resonant circuit in which high "Q" means it is able to store energy for a longer time. Switching "Q" meant varying the circuit's quality to store the energy or let it out. As their discussion continued, Culver learned that Gould had anticipated every one of his ideas and in some cases, as with the Q-switch, had gone beyond what he had thought of. Furthermore, Gould wasn't just thinking about an isolated device, but about the things that could be done with it.

"I've been looking for ideas like these," Culver said. "I see that somebody's figured them out."

The next day, April 1, Culver was pursuing his informal explorations of the laser in a meeting with Townes. They were talking in his office at Pupin when the phone rang, and Townes excused himself to talk. He talked for a few moments, nodding thoughtfully. When he hung up the phone, he told Culver that it was ARPA, telling him that the agency had decided to classify the TRG proposal. Culver had a copy with him. Its newly "secret" designation meant Culver could no longer carry it without a courier letter.

Townes told Culver, "I'll take it, and give it to the security office here at Columbia. They can send it on to you at Santa Monica."

Culver refused. He knew Townes had seen the proposal months earlier, and simply wanted to follow security regulations to the letter, but he was reluctant to hand over Gould's far-reaching ideas to Townes or anybody else. He took the proposal with him, absent the courier letter, and carried it back to Santa Monica.

■■

Back at TRG, the patent application was virtually complete when Goldmuntz learned that the proposal had been classified. ARPA had hinted that the pending contract would be classified as well, because of the military uses to be studied. That meant the patent application also was likely to be classified.

Keegan, consulting with Gould and Goldmuntz, but with all three of them flying virtually blind, took a guess at the information in the volume of material that might avoid classification. He extracted thirty pages, and put it into a smaller application. Outside the United States, Gould could file a classified application only in Australia, Canada, and the United Kingdom, the closest American allies. This would allow him to file in other industrial countries.

Gould wrote a $200 personal check to help pay for the preparation of the applications. Goldmuntz sent a secretary from TRG to help Darby and Darby's secretarial staff finish typing the applications on Saturday, April 4. A courier rushed them to the main New York Post Office on Eighth Avenue the same day, and they were received at the Patent and Trademark Office the following Monday, April 6, and assigned Serial Numbers 804,539 and 804,540, respectively.

The main application, Serial Number 804,540, was a thing of beauty. It laid out what were actually a number of inventions, all incorporated under the general term "laser." At its heart was the Fabry-Perot cavity with mirrors at each end—the "optically flat, partially reflecting parallel mirrors" Gould had conceived of eighteen months earlier—that would allow the light waves to oscillate back and forth and build in intensity until they burst through the partially reflecting mirror in a pure, straight, brilliant beam.

But the oscillator was only part of the laser. First the material, usually a gas, had to be excited, the beginning of the process that amplified the light. The top-heavy population of excited atoms had to be created, and the excited atoms had to start to emit their radiation before it could be steered between the mirrors and formed into a coherent beam. Gould's application set out two amplifying methods—optical pumping, and collisions of the second kind in a gas discharge—and identified a number of paired materials whose corresponding spectral lines would allow the atoms of one material to selectively excite the atoms of the other. This was the process that made the amplifier work.

The application also disclosed his plan for producing a beam with an enormous burst of energy. This was the Q-switch he had described to Culver. It revealed another innovation that was solely Gould's—the Brewster's angle window. A light beam projected through glass or any other clear material loses much of its intensity due to reflection, unless it enters at a certain angle—known as Brewster's angle—to the material. The angle varies with the material used; for glass, for example, it is about fifty-five degrees from vertical. Then, one polarization of the light streams through with virtually no loss of intensity. Gould had figured out that incorporating Brewster's angle windows in the laser would eliminate reflection losses, allowing the desirable polarization to build, and the other polarization to disperse.

Finally, the patent application identified the things Gould believed the laser could be used for, including manufacturing, heating materials to trigger chemical reactions, distance measuring, communications, television, and radar.

The application's specificity, breadth, and detail grew from Gould's conviction that in the laser, he had invented something important that promised much in the way of future royalties.

The ARPA contract, dated May 14, arrived six weeks later. It was, as expected, classified. The classification status of contract elements was laid out on the last page. Status and progress reports, inspection reports, and engineering notes and computations would fall under the less onerous "confidential" heading. But the essential elements—technical reports, tests, military performance or operating characteristics, and estimates or conclusions about the laser's capabilities or limitations—were designated "secret."

■■

Culver's trip to New York had brought an invitation to Gould and Goldmuntz to meet with Rand scientists in California. They were in Santa Monica, talking about laser applications with Culver and several others in a conference room when Culver's boss burst through the door. He snatched up Culver's copy of the TRG proposal and exited the room. From down the hall came the sounds of a furious discussion. Culver went to see what was going on.

"You can't be having this discussion," Rand's security officer told him heatedly. "Gould didn't send us his security clearance."

"But it's his work," Culver said.

"It doesn't matter. It's classified, and you can't discuss it with him."

Culver thought it was a mistake.

By then, TRG had begun its push to get Gould cleared. Nothing Gould had said to Goldmuntz made him think the problem was a big one. The excesses of McCarthyism had waned with public disapproval. McCarthy, now dead, had been discredited, and the McCarran Act, pushed into law by the House Un-American Activities Committee at the height of McCarthy's red-scare tactics, had been successfully challenged in the courts. Its provisions that U.S. communists be required to register as foreign agents, and denying them passports, were thrown out. But the act's ban on communists in government, and in defense industry jobs, remained intact.

With Paul Adams promising to do what he could to help expedite Gould's clearance, Goldmuntz contacted an old friend. He had known Adam Yarmolinsky since childhood, when they both attended the Midtown School, the Ethical Culture Society's elementary school in Manhattan. They both went on to the society's high school, the Fieldston School, as well. Yarmolinsky had obtained his law degree and was practicing in Washington.

Yarmolinsky had a solo practice, and was working primarily for foundations when Goldmuntz got in touch with him. Goldmuntz briefed him on TRG's pending contract and Gould's background. Yarmolinsky agreed to try to get Gould cleared, and brought in another lawyer, Harold Leventhal.

Goldmuntz took his plan to TRG's board. When the other board members asked how much it was going to cost, Goldmuntz bristled. "Forget about it," he said. "You don't worry about how much the doctor costs when the baby gets sick."

Security clearance applications asked the obvious question: Has applicant been granted/denied a security clearance in the past? Gould, having been cleared to work on the Manhattan Project and then fired, answered "Yes" to both. The few blank lines left for an explanation would hardly suffice in Gould's case. TRG's security secretary typed "See rider" in the space.

The rider was to be Gould's mea culpa.

Yarmolinsky and Leventhal began combing Gould's background and interviewing potential references. They looked for unassailable types. These included Gould's longtime friend and former Scarsdale neighbor Spencer Vickrey, whose father, Yarmolinsky wrote in notes he forwarded to Leventhal, "is in Who's Who, and is staunchly Republican, was active in relief work in

World War I, and was president of the Golden Rule Foundation for many years." He noted Polykarp Kusch's "granite-like and powerful personality" and Nobel Prize.

Gould had listed Alan Berman, his former laboratory partner, as a reference. He and Ruth continued to see Berman and his wife, Gwen, over bridge or at parties several times a year. But while Yarmolinsky noted that Berman was "married, 2 children. Dignified appearance. Very solid citizen in manner," he felt it necessary to add that on the down side, Berman wore a beard. The same was true of Hugh Byfield, a physicist who had studied at Columbia with Gould. Byfield at least had an excuse. His beard, Yarmolinsky noted, he wore "because a yachtsman."

Such distinctions revealed, even in McCarthy's aftermath, the temper of the times. Yarmolinsky noted to Leventhal that he had decided to omit as a reference the pastor of the Community Church in White Plains who had officiated at Gould's first wedding "in view of his free-thinking type of church, and Gould's lack of religious affiliation."

During the heart of the summer of 1959, the two lawyers interviewed Gould, his parents, his brothers and their wives, his aunt and foster sister, friends, and academic and scientific contacts. They visited the recesses of Gould's past. His youthful idealism and his many indiscretions loomed large, his love lives were dissected, family arguments and estrangements made their way from memory onto paper, and then into sober analysis of how they shaped the picture of Gould that the authorities should see. A constant thread ran through it all. Sturdy, solid patriotism was the ideal presentation. Gould's life was tough to work with, but from his palette of wild, dark, and sometimes radical colors, the lawyers would try to paint a Norman Rockwell portrait.

17

In all of Adam Yarmolinsky's interviews, Gould had emerged not as a communist but an enthusiastic capitalist along the lines of his hero, Thomas Edison. "The most important thing," Gould had told the lawyer, "is that under socialism you can't ever make a million dollars."

An even stronger recommendation, though it could only be inferred, came from Glen's flat refusal to cooperate with her ex-husband's effort to win a security clearance. Gould talked with her on June 24. He made notes immediately afterward and passed them along to Yarmolinsky. Glen, he wrote, considered him an enemy of the people and no longer wanted anything to do with him. Then working as a book editor, she spurned the notion of a "financial guarantee" that Gould had suggested might be possible. She had talked with a lawyer, and said that if any trouble arose for her out of Gould's effort to achieve a clearance, she would do her best to make trouble in return.

Gould summarized her response in a single bureaucratic sentence. "Glen's total attitude is non-cooperative."

With no obvious minefields in the information they collected from Gould's family and associates, the biggest problem Leventhal and Yarmolinsky faced was sifting through Gould's own version of events. He was reluctant to be entirely forthcoming, and pinning him down was like trying to grab smoke rings. Yarmolinsky's interviews produced several drafts of the explanatory rider. The drafts shuttled between Gould and the lawyers, the lawyers always wanting more details. Gould had said that while other members of the Prenski study group accepted Marxist theory, he "had reservations." When the draft landed back in his lap again, a note from the lawyers was attached: "G. to add specifics as to reservations."

The rider was to be a confession and a plea for mercy. Gould was supposed to admit youthful error and look back with chagrin at his foolishness. This was a humiliating exercise that he despised. It rummaged through his political dirty linen, and meant downplaying and even denying things he cared about. Gould's instinct for the underdog, his concern about unrestrained corporate power, his sense that all workers should be able to expect fairness and equal treatment remained strong, but to admit to these feelings would imply lingering communist sympathies. In the view of the lawyers, Gould had to throw himself at the feet of the Industrial Personnel Security Review Board. He had to make a clean breast of things, and convince the board that he not only no longer held the views he once had held, but now believed that they were all wrong.

The rider also had to anticipate what was in his dossier, and address every area in which the security agencies were likely to have information. An omission or half-truth would be as damaging as an outright lie.

Finally the lawyers reached a point of satisfaction with the information that they had. Gould's appeal began, "It was through Glen Fulwider, my first wife, that I was brought into contact with organizations listed in Form DD 48-1. . . . Through Glen I was introduced and, as I shall subsequently relate, became a member first of a Marxist study group; then the Communist Political Association; and then the Communist Party. I broke away from the Party group over a period of time and since my separation of Glen in 1950 have had no contact with the Communist Party or any other of the organizations listed in this questionnaire."

The rider continued to tell the story in Gould's voice. After twenty pages, he reached the point where, after he had put communism behind him, he faced the enforcers of the Feinberg Law. In sterile legalese, he referred to "an occasion in 1954 when a special sub-committee of the New York City Board of Higher Education called me up for questioning, apparently on my membership in the Prenski study group and the Sixth Avenue Club." On that occasion, Gould confessed, "Although I verified my own involvement, I found that I could not bring myself at that time to spell out the facts in detail."

Gould reiterated that he was told at the time of his appearance that the information he gave would go no further. The belief that his answers would stay within the hearing room was, in retrospect, naive in the extreme.

He went on to say, "I recognize that in view of my past background, my case cannot be an easy one for clearance." He said he had abandoned a passport application in 1956 because he expected it to be denied, and turned down a job offer from Hudson Laboratories that involved classified work because he knew he couldn't get a clearance.

"Now I feel that more is involved than foregoing a pleasant vacation, or even a good job. Now I feel that I do not have the right to withhold from the government my special knowledge. I have been fortunate enough to do some pioneering work in the field of missile detection and interception. I think I have a feel for the field, based in part on my years of pondering and research, that will be fruitful in further developments and applications. To some extent, of course, I have a financial stake in getting a security clearance. But I have had that before. What tips the scales now is the belief that I have an important and perhaps unique contribution to make to the safety and security of the country. I have therefore decided to put my case before the authorities charged with security clearance.

"I am clear in my mind that I am a completely loyal citizen. My intellectual and emotional attachments are now *anti* not *pro* the communist conspiracy and ideology. I have come to repent the foolish mistakes which started 15 years ago, when we stood with an entirely different attitude towards Russia. The events of the intervening years have opened my eyes.

"I believe that I should be held to meet the security standards in force on the work which I have pioneered."

■

When these final words were in place, Harold Leventhal wrote Larry Gold-muntz on July 9, "The rider is at the secretarial service and—I pray—will be ready to mail out tonight to Gordon Gould. Adam and I will read it over tomorrow morning—the day you receive this. If we all agree on the go-ahead sign, filing will be possible the same day."

Leventhal visited the Pentagon for a series of conferences on Gould's application eight days later. It was July 17, Gould's birthday. The lawyer met with a Colonel Orth and a Mr. Plant, and came away feeling cautiously optimistic. The review board that would decide the case, Leventhal wrote in a memorandum for the file, "at least is sophisticated and understanding as to former members," and understands "the disinclination to tell about former members" although by this time "there is little the government doesn't know," and naming names is "a test of good faith." It was essential, he wrote, that the person seeking clearance could demonstrate *"absolutely* no questionable activity since a certain past date," and that he cooperate fully with the FBI, since a damaging or questioning report "is fatal."

TRG had initially hoped that all components of the laser contract might be downgraded from "secret" to "confidential." But ARPA considered the laser too vital. Paul Adams had told Leventhal that "it is Gould's ingenuity in specific techniques that is critical, and that would necessarily be classified." In other words, Gould's potential contributions were in major, all-or-nothing areas. Adams also had discouraged TRG from seeking an interim clearance for Gould. The clearance Gould needed, to make what Adams called "a substantial contribution to a project of utmost importance," was a permanent one.

ARPA was content for the short term to have Gould out recruiting other scientists to work on the contract. Once his clearance was decided—the sooner the better, Adams had said—the agency was ready to "go full speed ahead—with Gould (we hope) or without Gould (if we must)."

■■

Leventhal's cautious confidence was shaken two weeks later, after Gould called him on July 27 to say he had run into Herbert J. Sandberg on a trip to Wright Field. Sandberg had taught physics at City College, Gould said, and he and Glen, and later he and Ruth, had spent time with Sandberg and his wife, Leda. Sandberg had gone on to Nordan Laboratories and won a security clearance after a considerable struggle. He now worked for Bulova Research in Queens. He could attest that Gould had left communism behind when he left Glen, and would cooperate to the fullest extent, Gould told Leventhal excitedly.

But Leventhal learned to his horror that Sandberg had used his friendship with Gould to try to deal with his own clearance problems.

The lawyer visited Sandberg just three days after he had heard from Gould. It emerged that Sandberg's difficulties had stemmed from his registration with the American Labor Party, and his membership in the Teachers Union. Both had been named by the House Un-American Activities Committee as communist-led organizations. These memberships led the Eastern Industrial Personnel Security Board to tell Sandberg in June 1954 it intended to deny him a clearance.

Sandberg had appealed. He led Leventhal down to his basement, where he fished out from some old papers the affidavit he had filed with his appeal. It accused Gould and Glen of suggesting he join both the ALP and the union. He had not suspected them of being communists at the time, but he did later, "particularly Mrs. Gould."

The Goulds figured heavily in Sandberg's appeal as he tried to play his suspicions about them to his advantage. His affidavit said Gould "kept hounding me" for Teachers Union dues. He produced from the box of papers a letter he had written to the FBI's New York office on August 22, 1951. It detailed a vacation Sandberg and his wife had taken in Provincetown, Massachusetts, during which they met, through his daughter, the economic editor for *Tass*. Having made this confession, he went on to ask the FBI's advice about how he should conduct his friendship with Gould.

"We know his beliefs are very liberal," he wrote, "but we are wondering whether he is a CP member. Although we enjoy his friendship there are some things that come before friendship. We would rather not see him anymore if their loyalty is doubtful. In times like these we believe there is no place for split loyalties.

"Although I have asked him if he is a party member, he would not answer either way, which makes me doubt even more."

The FBI responded with Orwellian doublespeak. Edward Schmidt, special agent in charge of the New York office, wrote that the information contained in the bureau's files was confidential. He added, "I feel certain that upon due reflection you will recognize the need for such regulation, and that my inability to be of assistance to you in the matter of your inquiry will not cause you to conclude that we do or do not possess the information requested."

Still later, Sandberg took another stab at using Gould, who had maintained the friendship after he and Glen had separated. He and Ruth had invited the Sandbergs on a sailing weekend, where they had again found matters of concern. Ruth had brought along some fish to cook, and when she opened the package Leda was dismayed to see it had been wrapped in the *Daily Worker*. And at some point during the weekend, Gould speculated that American forces might have used germ warfare in Korea. Some time later, when Sandberg was visited by an agent with regard to his own clearance, he fingered Gould.

Sandberg's affidavit said, "My wife and I offered our services to Mr. Jim Rogers of the Federal Bureau of Investigation on the occasion of his visits to us in July 1953 in regards to Mr. Gould. My wife and I have always been willing and anxious to give our aid in any possible way to help destroy the communist menace both here in the United States and throughout the world."

Leventhal could only guess that since the Sandbergs had offered to report on Gould to the FBI, they might have spoken about the sailing weekend to the agent. He feared that somewhere in the FBI's files was innuendo—a fish wrapper and a casual comment, but that was all it would take—that contradicted Gould's contention that he had finished with communism when he left Glen.

■

In August, the case remained unresolved, and Goldmuntz was concerned about the mounting legal bills. The baby's extended sickness had him worrying about the cost of the doctor after all. When he demanded more detailed accounting from Leventhal and Yarmolinsky, Leventhal wrote back on August 18 to say, "This case has come to require more services than I originally anticipated.

"This is largely due to the fact that in successive interviews it has emerged that Mr. Gould has a less distinct memory than is customary concerning the critical period involved." This had meant more revisions and redrafts of the rider. "That the case is not a conventional one, and shadings are important, has underscored the importance of accuracy in our submissions," Leventhal wrote.

He went on to describe the difficulties presented by the case, and the need to see what charges the government's files on Gould contained "to judge what lies ahead." He anticipated that he would need one or two conferences with Gould to prepare for the interviews he expected Gould to have with security clearance personnel.

In September, Gould handed Leventhal a potentially helpful "shading." He remembered that he had refused to sign a petition circulated by the chemist Linus Pauling calling for a suspension of atom bomb testing.

"I take my hat off," Leventhal applauded in a letter.

Gould by then was chafing at the seemingly endless delays. Leventhal wrote him on September 24 to counsel patience. "I am, believe me, fully aware of how much you must be, as you say, looking forward to the end of this dreary business. But the situation is that now that the request has been made for expedited handling of the case, and now that we know that the investigation is in fact underway, there is literally nothing that can be done at this time." He didn't want to urge the intelligence authorities to greater speed, as that might "border on complaint of the way intelligence is doing its job," and told Gould to "grin and bear it."

Shadings continued to worry Leventhal. Their subtlety signified just how deep paranoia had seeped into the nation's bones. He wrote in the same September 24 letter, "It seems to me that it was not until 1957 that you felt clear enough about your convictions, or about the fact that they would continue in effect, that you decided to register as a Democrat." He fretted that the authorities might tie this new registration to Adlai Stevenson's call in 1956 for a unilateral suspension of atomic tests, and told Gould, "I do hope that you can find someone to corroborate the fact that you did refuse to sign the Pauling petition. I regard it as extremely significant."

When he was finished with the letter, Leventhal dictated a note to the file. Gould would have to make a statement when he appeared, and Leventhal wanted him to be familiar with all the hot-button issues on which the authorities judged communist sympathies. The lawyer called his list "Touchstones and Shibboleths" and catalogued several areas where he knew Gould would have to have the right responses. Did Gould really believe American forces had used germ warfare in Korea? How did he feel about unilateral nuclear disarmament? Was Russia being encircled, and should America withdraw from Europe? He also listed the 1952 trials of Alger Hiss, the trial of Julius and Ethel Rosenberg, Congressional investigations of communism, and Italy's possible fall to communist control as potential subjects.

18

In September 1959, with Gould preoccupied with the "dreary business" of his clearance, Charles Townes brought together the first International Conference on Quantum Electronics. The rubric was a new one, coined from electronics and quantum mechanics. The Office of Naval Research sponsored the conference after its project monitor at Columbia, Dr. Irving Rowe, suggested the idea to Townes. Scientists from around the world converged on Shawanga Lodge, a resort hotel in the Catskill Mountains town of High View, New York.

Townes, having seen TRG's proposal, certainly knew by then of Gould's patent application. He apparently viewed Gould as no threat to his own patent claims, however. Art Schawlow was chairing a session on optical masers, and Townes, reasoning that TRG was a new company trying to get started in the field and Gould "was a young man trying to do things," urged his brother-in-law to "say something good" about Gould.

Schawlow complied, but only grudgingly. For one thing, he was annoyed that TRG's contract and Gould's primary patent application had been classified. He suspected that TRG was trying to persuade the Air Force to classify all laser research, and feared that Bell Labs' work would also fall under military control and secrecy rules, which created annoying data handling problems. For another, the Russian scientists Basov and Prokhorov were attending the conference and he didn't want to "air our dirty linen in front of the Russians."

A genuinely nice man, not much given to dwelling on personal antipathies, Schawlow nevertheless thought back two months to a conference he'd attended at the University of Michigan. Peter Franken, Gould's former laboratory partner at Columbia, had organized the conference, the first in the world on optical pumping. He had invited Gould to give a talk, in which Gould cheekily kept referring to the "laser" rather than the "optical maser." This was an affront, as if he'd invented the word. Schawlow had been annoyed about that, too.

Further, Gould had talked about the many possibilities he saw for making a laser work, but didn't describe them. It seemed to Schawlow that Gould had said he had devised several ways of exciting a laser, and several structures, but would say no more since they were classified. He did, however, suggest that a nonresonant cavity with white walls instead of mirrors might scatter radiation back on itself and produce amplification. This would create a very narrow frequency standard independent of cavity length. Schawlow hadn't heard this before, and in his talk at Shawanga Lodge threw Gould a bone by crediting him with the idea.

But he couldn't resist also bringing his acerbic wit to bear on Gould for his temerity in nomenclature. "It seems to me that your 'laser' would be more useful as an oscillator than an amplifier," Schawlow remarked. "So if we substitute an 'o' for the 'a,' your entry in the optical maser race really should be called the 'loser.' "

Schawlow complained to Townes later for pushing him to mention Gould. "I wish you hadn't insisted on helping Gould out," he said, "because that really was not much good. I tried to do the best that I could for him."

Gould was already being dismissed, the science elite lining up against him. It found his emphasis on applications unworthy, the pursuit of a tinkerer. His instinct to put his patent rights first, and sharing his ideas second, was unscientific. And his refusal to play follow the leader was not only independent, but arrogant. All this from a man who, having failed to complete his Ph.D., could not be a serious person.

Schawlow's slight to Gould was forgotten by almost everyone. But another remark he made at Shawanga Lodge was remembered. As scientists pondered the kinds of materials that could be used in a laser, one of those they focused on was ruby. They considered not a gemstone, but a tube-shaped, artificially produced pink ruby, whose chromium content, when excited, produced resonance lines that offered possibilities. Schawlow, however, referred to measurements showing that the absorption lines and the associated fluorescent lines that might produce laser action had too low a quantum efficiency. Too much energy was lost in the process, he said, and therefore ruby was unsuitable for a solid-state "optical maser."

One of the physicists in the audience, Theodore H. Maiman from Hughes Aircraft's Research Laboratories in Malibu, California, noted Schawlow's figures with interest. They disagreed with his own calculations. Afterward, he went home to California, rechecked his figures, and decided to continue working with ruby. Others, taking Schawlow's word as gospel, abandoned ruby as a laser possibility.

■■

The fall of 1959 deepened into winter. Almost three months passed. The New York office of the Industrial Personnel Security Review Board approved Gould's clearance application. But Washington had the final word, and it reversed the New York decision pending an interview with Gould. The flood of memoranda, interview notes, and correspondence going into Leventhal's and Yarmolinsky's TRG files slowed to a trickle. Meanwhile, agents of the Office of Special Investigations set up an interview, then postponed it. It had not been rescheduled on December 16, when Leventhal sent Gould his list of "touchstone" issues.

He had added some components to his litmus test of Americanism in the late 1950s. The new issues included Soviet intervention in Hungary, Tibet, and Laos, Russia's history in Finland, and the Yalta conference at which Roosevelt, Churchill, and Stalin had argued for their spheres of influence at the end of World War II.

Leventhal had interviewed Gould extensively about all these things, but he continued pressing to refine Gould's answers. Gould, in his conversations with the lawyer, had been unable to conceal his disgust at the excesses of the anticommunist crusade. He may have remembered events imperfectly, or not at all, but he was clear and direct about how he felt about McCarthyism, and he seemed unable, or unwilling, to temper his opinions. Leventhal wanted him to be more circumspect. He offered some lawyerly coaching. "I gather that you feel a little more clearly that there were abuses in the Congressional investigation of communists, but that it is not easy to avoid these abuses when the Committee is concerned with awakening the American public to the danger."

He urged Gould to get a copy of J. Edgar Hoover's *Masters of Deceit* as a guide to how Gould might have been unwittingly manipulated. "There may be some passages which will recall things to you," he urged, "some instances that he gives of communist techniques which may recall incidents that are now buried."

■■

TRG, meanwhile, had no choice but to start working on the laser contract without Gould. Research labs across the country were gearing up experiments, and the race was on. Word had spread about the laser applications Gould had suggested in TRG's proposal, but nobody was trying to create those kinds of applications. Hardly anyone even believed they were possible. The race was one for prestige, and a place in scientific history, for being the first to prove simply that light waves could be amplified.

Large doses of swagger issued from some corners. Word reached Larry Goldmuntz that dedicated Bell Labsters were making sport of TRG and calling the Long Island firm "Pipsqueak, Inc."

The company was certainly smaller than the research behemoths it was up against, Bell Labs included. But it was not as weak as the diminutive appellation would imply. TRG had expanded, modestly but steadily, since its beginning. When TRG won the laser contract it had already advanced the design of atomic frequency standards by making them smaller and lighter under its contract with the Signal Corps. It was developing new navigational equipment and systems for aircraft and missiles, building microwave components and antennas, and radomes. It was doing theoretical work in hydrodynamics aimed at designing slippery ships that could pass through water with less wave resistance. TRG had designed the radio feed for the Arecibo radio telescope in Puerto Rico, the world's largest. The company continued to work in military systems development and nuclear reactor and shielding research, and on new heat-resistant materials for rocket nozzles and radomes. TRG had busy electronics and tube laboratories, a machine shop for prototype production, and a computer center where an ALWAC III-E with an 8,192-byte memory hummed along on vacuum tubes as its operator changed programs by moving plugs on a board that resembled a bygone telephone exchange.

The laser contract had accelerated the pace of expansion drastically. TRG had spilled into the building next door. Four Aerial Way was a twin to No. 2, a low, sprawling building of red and white bricks containing about 15,000 square feet of work space. The lunch-hour athletes who had played football on the roof at 17 Union Square now strung a net between the closely spaced buildings and punched a volleyball back and forth. TRG installed all its classified activities in the new building, and controlled entrance and egress according to security requirements.

Gould was relegated to a windowless office at the core of No. 2, where he had to use a separate entrance and pass by the security secretary's desk on the way to his door.

Dick Daly's department—now called the Quantum Electronics Department—moved to No. 4, and shifted its primary focus to the laser. After originally resisting the very idea of the laser contract, Daly now headed a group of perhaps twenty men and one woman, backed by experts in optics and spectroscopy, technicians, a metallurgist, and glassblowers to fashion lamps and beam tubes, whose mission was to build the groundbreaking device.

It was not a good arrangement.

Gould's absence from the experimental team was the biggest problem imposed on TRG, but not the only one. Daly didn't like Gould. You could speculate all day about the reasons—their politics and lifestyles clashed; Daly's expertise had taken second place to Gould's; Gould lacked a Ph.D.—but the effect of Daly's disaffection was to undermine even the contributions that Gould would have been able to make under his restricted status.

Daly was one of those people who had to be the best at everything he did. A bomber pilot during World War II who maintained his private pilot's license, when one of the technicians at TRG earned a commercial flying license, Daly went out and got his, too. He bought a hunting bow with a fifty-five-pound pull, quickly tired of it, and placed a notice on the company bulletin board advertising it for sale; asked why he was selling it, Daly responded, "It's too weak." He was getting an eighty-pound bow, he said, one that required a great deal more strength to use. Lunch hours at TRG featured floating bridge games, and not infrequent hints that Daly had misplayed his partners' bids, to which he would thunder in defense of his superiority, "I can swim, I can ski, I can sail, I can fly an airplane, and I can play bridge."

His own ideas were, of course, better than anybody else's. So when an experimental choice came down to Gould's versus Daly's, Daly's got tried first. Daly encouraged skepticism toward Gould's ideas among the doctorate-level physicists who were working on the laser, even the ones Gould had brought to TRG. Ben Senitzky was one of Gould's hires. He liked to say, "You can't buck Boltzmann," implying it would be impossible to create the population of excited atoms necessary to create a laser. Daly fed this sense that if their experiments were to prove anything, it was that Gould's vision was nothing but a pipe dream.

"The Ph.D.s just didn't have much technical respect for Gordon," physicist Paul Rabinowitz recalled. "They looked down their noses at him, because he had never completed his thesis, and wasn't a well-known scientist. He was not a theoretician, nor a great mathematician. A lot of his scientific instincts were intuitive. The ideas he had about the laser seemed Buck Rogerish. But his ideas weren't just wishes. He had good technical reasoning, and he could explain why certain things could work. Their attitude was the opposite of what it should have been."

The fact that Gould was smarter than Daly, and that Daly almost certainly realized it, made matters worse.

"Daly was in charge of the group, and the people worked for him," said Ted Shultz, an optical scientist who started work at TRG in 1959. "But Gordon had all the ideas, so people were always talking to Gordon."

Under Gould's security restrictions, what they could say to him was governed by a set of rules that could only be termed bizarre. They could ask him questions, as long as they didn't provide him with enough information to know what they were doing.

■■

The experimenters worked in several teams, each investigating one of the laser types Gould had proposed. One was ruby. Another team worked with optically pumped alkali vapors, starting with potassium. A third approach was Gould's gas discharge laser using collisions of the second kind, employing a mixture of krypton and mercury gases in which metastable krypton atoms transferred their energy to atoms of mercury.

Daly, while he headed the whole group, took personal charge of the experimental team working with ruby. Art Schawlow had taken some new measurements and was no longer damning ruby as a laser substance. In fact, many physicists now considered optically pumped ruby the best possibility for making a solid-state laser.

Gould was no expert in condensed substances like ruby, and it was also one of the areas in which his access to detailed information from the experiments was restricted since he'd proposed military applications for ruby lasers. The same was true of the krypton-mercury collisions scheme that he alone had proposed; it was in the classified version of his patent application. Gould's greatest knowledge, growing out of his work with thallium, was about optically pumped alkali gas lasers. He'd proposed potassium, as had Schawlow and Townes, and also cesium. Both media were listed in Gould's unclassified patent application, and so fell outside the worst aspects of the ban imposed on him.

Rabinowitz and another young physicist, Steve Jacobs, got the assignment to investigate the alkali gas laser, and Gould could actually talk to them, although under tight restrictions.

■■

Rabinowitz, tall and somewhat sedate in manner, had grown up in Bensonhurst, a Brooklyn neighborhood of middle-class Jewish and Italian families, and majored in physics at Brooklyn College. He worked there after graduation, setting up laboratory experiments for undergraduates while he studied for a Ph.D. at New York University. Then his wife told him she was pregnant. Facing the need to make real money in the middle of what was being called the Eisenhower Recession, Rabinowitz interviewed at TRG.

Gould, who led the interview, liked the younger man. He sensed in Rabinowitz an intuition similar to his own, and he didn't care that he lacked a Ph.D. TRG dangled a magnificent salary of $6,500 in front of him, and Rabinowitz started early in 1959.

Jacobs followed Rabinowitz by a year, after TRG had started working on the laser contract. Gould hired him at the New York meeting of the American Physical Society in late January 1960, after Jacobs described his undergraduate studies at Antioch College in Yellow Springs, Ohio, and his Ph.D. in spectroscopy from Johns Hopkins University. Jacobs would always remember the way Gould's eyes danced when he talked about the laser, and a month later he had quit his job at the Perkin-Elmer Corporation in Norwalk, Connecticut, where he was working on space optics projects, and moved to TRG.

Rabinowitz was working on a cesium vapor laser optically pumped by a helium lamp when Jacobs arrived. Potassium hadn't shown any promise, and then Rabinowitz had caught an electrical shock in the laboratory that almost killed him. Recuperating in the hospital, he was browsing through reading material only a physicist could love, Mitchell and Zumansky's *Atoms and Resonance Radiation*. He read of an experiment in which cesium was pumped by helium radiation, which Rabinowitz knew was one of the laser schemes Gould's proposal had suggested, and he went to Daly to suggest the switch.

Daly had resisted, until he learned that researchers at Columbia had had the same insight. Herman Z. Cummins, a graduate student working under Townes, had taken on the "optical maser" as a thesis project. Townes was on a leave of absence from Columbia; he had moved to Washington as vice president and scientific director of the Institute for Defense Analysis, but he continued to work with graduate students and checked in on them every week or two. He and Cummins also had decided to give up on potassium, and Cummins was now working with helium-pumped cesium.

Jacobs, wiry and energetic, fit in well with Rabinowitz's more laid-back manner. Together, they personified TRG's shoot from the hip style. Jacobs, the more organized of the pair, would lay out work schedules and routines. Rabinowitz shared with Gould an experimental intuition that allowed him to make on-the-spot adjustments.

Most important, the two of them were believers. Rabinowitz was too young to be hampered by doubts, and Jacobs was swept away by Gould's zest for his invention. They managed to ignore TRG's internal politics. They were happy to be Gould's acolytes, and they worked hard to please him.

But the rules that governed Gould were such that while the two could come to him with questions, and even let him know what they were doing, he could not enter the laboratory or take part in their experiments. This approach reduced Gould to a theoretician. Mathematical theory was Gould at his weakest. He was a back-of-the-envelope physicist, who let the numbers follow his sense of what would work. To be confined to slide rule and paper was, for Gould, like being shackled to a ball and chain.

19

Frustration gnawed at Gould as the work went on around him. His role in recruiting a crack team of physicists to build a laser, which had continued throughout 1959 and into 1960, was over now that the team was in place. It had been easy to share his enthusiasm for the laser's possibilities while he was recruiting. Now, barred from learning the details of their experimental work, he would pass the same physicists in TRG's common areas and in the parking lot and see their looks that verged on pity.

Gould could enter No. 4, but only as far as the conference room. Most of the people working on the laser brown-bagged their lunches there or at their desks. This was different from the days after TRG had moved to Syosset, before the laser contract, when twenty scientists with protectors in their shirt pockets and slide rules on their belts would descend upon local restaurants for lunch and talk of atomic physics. Gould played bridge across the conference table with Daly and some of the others. When lunch was over, and everyone else had gone back to the classified laboratory, Gould returned to his isolated cubbyhole.

The situation was absolutely nuts. To get to the men's room, Gould had to pass through a classified area. Goldmuntz had had to call in a contractor to knock down a wall and build another door to the facility.

Visiting dignitaries kept dropping in, making Gould's situation even weirder. Teams from the Air Force Office of Scientific Research and elsewhere in the Pentagon, famous scientists from other institutions, and oversight committees from Congress and the military were always trailing escorts through the TRG labs and sitting down to talk with the scientists working on the laser. Gould's proposal had touched off the most amazing science fiction fantasies, of light rays knocking missiles out of the sky. The Air Force was calling TRG's research Project DEFENDER, and the visitors all were curious about what this incredible light ray would be like, and what could eventually be done with it. They dutifully met Gould as the proposal's author. Then, when the talk turned to experimental gains, Gould had to excuse himself.

Safely in his office, Gould spent his time studying the Table of Elements, searching for matching spectral lines that offered new—and unclassified— laser possibilities. He had been forced to surrender his eighty-seven-page notebook on which the proposal had been based. It was locked in a safe as soon as the project was classified. Gould, anticipating the classification, had made copies, but since it was illegal for him to refer to his own work he couldn't do so openly. He had plenty of other things to work on, unclassified projects, but much of the time he sat in his isolation, pondering questions that were phrased in such a way as to prevent him from divining their context.

On March 22, 1960, the Patent Office granted Arthur Schawlow and Charles Townes patent number 2,929,922 for "masers and maser communication systems" based on their optical maser application. When Gould read the news in the inside pages of *The New York Times,* his frustration ratcheted upward at being restricted from the lab.

■■

The week the Schawlow and Townes patent was awarded, the Office of Industrial Personnel Security Review finally invited Gould to appear before its screening board. Leventhal called the Pentagon on March 28 to accept the invitation. A return letter from a Herbert Lewis in the director's office confirmed an April 19 date for Gould's interview. He was instructed to appear at 10 A.M. in room 4B943, the Pentagon.

Lewis's letter to Leventhal was dated April 1, April Fools' Day, Gould observed with gallows humor. He thought his chances were dim. The shadow of McCarthyism still lay over the land, and he simply did not believe his appearing before the screening board would change things. But he made plans to attend.

Gould caught an early flight from LaGuardia Airport to Washington National for the Tuesday "interview." He took a cab to the Pentagon, where he met Leventhal and Yarmolinsky. Together, Gould and the two lawyers found room 4B943. In the interview room, a Captain Davis sat at one end of a long table where several other men were seated. The interview got under way, and Davis guided the panel through a series of questions directed at Gould.

Gould answered impassively, trying to contain his temper and keep Leventhal's "touchstones and shibboleths" in mind. A stenographer tapped away at the makings of a transcript as Gould patiently and calmly recounted the events he had gone over time and time again. He remembered to mute his criticism of the Congressional overreaching that had made the crusade against communism into a witch-hunt. He offered to provide affidavits from people who had known him since 1950, who could say he was not "communistic" in his thinking, and from people he had known since the 1940s who could attest to his shift from a "left-winger" to a right-thinking American. At the end of his presentation, Gould tried to read the faces of the panel and felt in his bones that nothing had changed.

Gould flew back to New York in a resentful funk. He had a thought and laughed at it bitterly. Communism's biggest problem, the reason it would never work, was that under communism people had no incentive to pursue work they loved. How ironic it was that in the name of anti-communism, he was barred from working on his passion.

His next letter from Leventhal was dated April 21. The lawyer said he had been advised by Captain Davis that the board would not rule until it received a transcript of the session. Davis had urged that Gould submit his affidavits quickly. He had said the board would be very happy to receive "any evidence of the transition of the thinking of Mr. Gould." From this Leventhal inferred that the most valuable affidavits would describe Gould emerging from the grip of communism.

Ruth Gould already had written hers. Her husband had returned from Washington the night of the hearing, told her what had happened, and described his feeling of futility. Ruth got up early the next morning and sat down at her typewriter. She supported her husband fiercely, and said so. "Since wives are biased by definition, and since I do not consider myself an exception, I will confine myself to the facts and leave the interpretation to you," she wrote.

She described how Gordon had written her at Yale after he had gone to work at the Manhattan Project: "He never explained what was going on. Only how exciting it was, how much he sweated, and how many salt tablets he had to eat."

Gould's post–Manhattan Project communism was Glen's "doings." It made no sense otherwise, Ruth wrote. "Gordon had never been vitally interested in politics—only mildly and good-naturedly so. Politics was an intellectual pursuit, and intellectual pursuits were important to his family, so Gordon was interested, too. But he was far more interested in physics, sailing, bridge, mountain climbing, choral singing and last, but very far from least, girls."

After the war, when he stopped by the Memorial Hospital Center where she then worked in order to give blood, and afterward visited with her, she recalled thinking, "My goodness, I heard he is a communist. Why doesn't he try to convert me?"

After they began to see each other steadily in the spring of 1952, he rarely mentioned communism, she wrote. She went on to quote Gould to the effect that he now felt worlds apart from the people he had met as a communist, that they were naive, and that he had nothing in common with them anymore.

Peter Franken also responded quickly. He sent Yarmolinsky a draft of several pages from which he suggested the lawyer construct an affidavit. Franken's rambling submission described a social friendship with Gould from their Columbia days that had little to do with politics. He spoke to Gould's integrity, but not his conversion. He also described Gould as a man at war with himself: "Gordon has been something of an 'accident prone' professionally. He is an extremely good scientist, creative and exceptionally endowed intellectually, but there appeared to be elements within his personality that precluded an effective exploitation of his talents. During the time I worked with him I had a strong sense that 1/3 of Gordon was working against the other 2/3 and that did not leave very much for conquering new horizons."

While he sensed that that had changed, Franken went on, "I think that if Gordon is thwarted at this, his first potentially successful and very useful effort, it will be just about the end of the fellow. I have the sense of a man of 40 or so who has been mustering his efforts for one big completely integrated attempt to function brilliantly and constructively. He has bet a large number of emotional cookies on this one and I would hope very much, with the national interest in mind, that he could push this project as far as possible."

Gould's brother David and David's wife, Toni, co-signed an affidavit that, like Ruth's, attributed Gould's communism to Glen's influence, which moderated as soon as they parted. "His present wife, Ruth, has a very stabilizing influence on him, and has done much to help him outgrow a certain immaturity," they wrote.

By the second week in May, Leventhal had submitted eight affidavits and received an acknowledgement of them from Captain Davis. Leventhal conveyed his impression to Gould that Davis was aware "that these letter-affidavits spring from the heart of the affiant, and are not merely work-products of your counsel."

The security review board took its time acting on the letters. Once again, Gould chafed. "It's difficult to be patient, but that's the course I counsel," Leventhal wrote to Gould on June 14.

■■

Three weeks later, news from California riveted the scientific world and made news broadcasts and front pages all across the country. The Hughes Aircraft researcher, Theodore Maiman, had made a laser work. The device was a rod of ruby not much bigger than a cigarette butt, silvered at the ends, triggered by a flash lamp that wound around it like a coil spring, and it created for a few millionths of a second a beam of light of the kind of power Gould had predicted—10,000 watts.

News of the first working laser shared space on the front page of the *Times* with reports of Senator John F. Kennedy's advance toward the Democratic presidential nomination. The story was agog with future possibilities, many of which Gould had predicted from the outset. A perfected laser, it said, would produce a narrow beam intense enough to illuminate "small swaths of the moon's surface or vaporize materials placed in its path."

The picture shown was not the laser that had worked. Hughes publicists deemed that device too small for something that was trumpeted as the next coming of Thor, and trotted out a larger, but nonworking, laser model at their triumphant news conference.

Maiman's breakthrough did not rate the pages of *Physical Review*, however. The scientific establishment, deferring to Townes and Schawlow, still had not embraced "laser" as the term of choice for the new machine. Maiman had submitted a concise, three-hundred-word article entitled "Optical Maser Action in Ruby" to the editors at the prestigious journal, only to have it rejected on grounds that advances in maser physics were no longer big news. The paper appeared, instead, in the British journal *Nature*.

Gould had now lost both of the contests that would have covered him in glory and ushered him automatically into the history books. Schawlow and Townes had the patent, and now Maiman had the first working laser, and the breathless news coverage that came with it.

Gould didn't care particulary about the glory. He was not fond of attention. He did care, however, about the substantial royalties he foresaw would be attached to an invention of the laser's significance. He knew in his gut he had thought of and described it first. He knew that no one had conceived the key to amplifying light and a light beam before he had, no one else had given it a name, no one had written anything valid that predated his 1957 notebook. He had a year from the date that Schawlow and Townes had received their patent to place his claim against theirs, and so he bided his time.

■■

The letters from the lawyers grew more vague about Gould's prospects. The board put off acting on Gould's case, pending the issuance of new security regulations governing defense contractors as a result of the successful challenges to the McCarran Act.

In August, Leventhal spent an hour and a half with Tyler Port, the administrative head of Industrial Personnel Security with the Department of Defense. Port had nothing to do with making a judgment on Gould's case, but Leventhal hoped he could expedite a hearing.

Port was not hopeful. Cases were backed up, he said. Many had gone on even longer than TRG's. Not only that, but the difficulty of securing witnesses, often many years after the events in question, made it difficult to schedule cases.

Leventhal argued that TRG was doing important work, and Gould was needed for it, but all he got was Port's promise to raise the question with the new chief counsel for the board.

Gould received his copy of Leventhal's account of the meeting stoically. Goldmuntz turned his hopes to the presidential election. The Democrats had nominated Kennedy at their convention in July, and Goldmuntz believed that if Kennedy was elected over the Republican nominee, Vice President Richard Nixon, in November, the last vestiges of McCarthyism might at last be buried.

■■

In the laboratory at Syosset, the switch to cesium produced a new set of experimental problems. Cesium, an alkali metal, is the most electropositive element known, meaning that it gives up its electrons easily. Cesium dropped into a glass of water will explode. A piece of cesium exposed to open air at room temperature will catch fire. Heated to a vapor, it will try to attack everything it touches. The first challenge for Jacobs and Rabinowitz was to find materials that would withstand it.

Pyrex was one. They sat down with Gould and worked out an experiment in which a ball of cesium in a Pyrex cell with triethylene glycol, the compound used in antifreeze, was heated to about 150 degrees centigrade. The cesium fluoresced. This told them they had the necessary population of excited atoms to produce a laser. Jacobs wrote up the results and, in September, sent a paper under all three names to the American Optical Society. He was convinced that he, Gould, and Rabinowitz were on the verge of building the first continuous wave laser.

But that was a more complicated experiment, one that compounded the difficulties cesium presented. They knew there was a transition between energy levels in cesium at a wavelength of 7 microns—or micrometers—and that was the laser Gould wanted to build. The problem again lay in finding materials.

The containers had to be cesium-proof, but also transparent for optical pumping. It was almost impossible to find something that was transparent at 7 microns, that cesium wouldn't eat. Glass wouldn't work. Neither would quartz. Neither would sapphire. The 7-micron transition was well into the infrared, where there were few detectors, and even fewer materials that would transmit it. Progress was agonizingly slow.

In TRG's tube laboratory, a tall, slender Greek-American named John Poulos designed and fabricated flash lamps, electron tubes, and other experimental equipment. Poulos was a man of infinite patience. He had worked at Western Electric and Sperry Rand before coming to TRG, and he prided himself on having never broken a piece of optics. But he bore the burden of "a lot of careless professors, who were a little sloppy."

Poulos's shop was two doors away from Gould's office in a corner of TRG's administration building. Rabinowitz and Jacobs kept Poulos, and his small staff of technicians and a glassblower, busy constantly. Exploding lamps took time to redesign and assemble according to new parameters. The laser itself started as a crystal-clear tube, aligned between curved reflectors designed to bounce light from the helium lamp onto the tiny cesium cell within. There were mirrors beyond either end of the cesium cell that had to be perfectly parallel to reflect the light back and forth, angled windows to reduce reflection, the whole affair bristling with protruding viewing ports and output barrels, ports for pumping off air to create the necessary vacuum, and electrodes for power input.

For Poulos and his crew, assembling the components was a task as delicate as building a tiny ship in a bottle. For Jacobs and Rabinowitz, performing the experiments was like balancing that ship on the head of a pin, and trying to keep it balanced while the glass of the bottle clouded to invisibility as the cesium attacked it.

It seemed to Gould that the two of them were at his doorstep constantly.

Normally he was patient, but he was feeling pressure. There was always some communication among the cutting edge laboratories, especially those in close proximity. It wasn't often official. Researchers at one facility would sometimes make presentations at another to talk about new developments, but usually it was a matter of a phone call here, a casual meeting there, at which physicists didn't so much share secrets as toss around ideas. In that way almost everybody knew what everybody else was doing, and if they didn't know for certain, they could guess. Townes and Schawlow weren't pursuing experiments themselves, but Gould knew Townes's student at Columbia was working with cesium, and he wanted to be successful first. Each time Jacobs and Rabinowitz came to him with some new technical problem to report, Gould bristled. "I know, I know, cesium presents big problems," he snapped. "Just go back and deal with them."

Gould was allowed to sit in on technical meetings, but when he spelled out experimental changes, his suggestions were routinely overruled by Dick Daly. Daly and the physicists working under him, early in the fall of 1960, had duplicated Ted Maiman's success by making a ruby laser work. This inflated Daly's already high opinion of himself and his engineering capabilities. Once they were back in the laboratory where Gould couldn't follow, Daly would tell Jacobs and Rabinowitz to scrap Gould's suggestions and follow the methods he proposed.

His attitude, according to Rabinowitz, was "Don't confuse me with the facts. I've made up my mind."

From the outset, however, Gould's array of insights into the laser had allowed TRG's scientists to deal successfully with a problem that vexed other laboratories as they conducted their early laser experiments. His Brewster's angle windows, the size of a little fingernail, tipped at the desired angle to the beam, were among the most delicate of the laser components. Poulos and his team used glass braising to fix them in place, and reflection losses were never a problem in the TRG experiments.

20

Gould's affiants had attributed to him a stable life, in contrast with the tumult of his bygone relationship with Glen, in their efforts to help him win a security clearance. In fact, Gould's personal life had returned to chaos.

His marriage with Ruth had been unhappy for some time. Their sex life, vigorous before the marriage, was almost nonexistent. Gould flayed himself with guilt for even marrying her, because he was now convinced that he hadn't loved her, and that his impulse to marry had grown out of the sense of obligation drummed into him by the straitlaced Methodists in his mother's family. His aunt Ethel was not a missionary like her sister Margaret, but she was every bit as stern. "Men are tomcats, despicable tomcats," Ethel liked to say. Poor Gould. He didn't want to be despicable, so there was only one interpretation of his aunt's invective—if you slept with a woman, you had to make an honest woman of her. Now he felt trapped. Ruth's selfless support and constant encouragement only deepened Gould's nagging guilt.

Another reason for guilt was his affair with Marcie Weiss.

It would have been an ordinary affair, except for one extraordinary thing. Even Gould had to admit the audacity of it. Marcie was TRG's security officer.

The affair grew from proximity as much as anything. Gould passed by Marcie's desk each day on his way into his isolated office, and they had fallen into long conversations as Gould smoked one cigarette after another. Marcie, a transplanted Texan, was everything Ruth wasn't—tall, statuesque, good-looking, and not a bit concerned with whether Gould ever got his Ph.D. She knew most of the details of Gould's life, having seen them in his submissions to the security authorities. She knew that he had suffered for his ideals, that he had carried on a doomed romance with a partner for whom he had sacrificed everything, who had then poisoned him politically. This was irresistibly romantic. She also knew he was smart, having almost single-handedly moved TRG into a phase of wholesale expansion, and at the same time he was vulnerable and wounded. An attraction developed.

Gould had a weakness for women who were attracted to him. He was attracted to their attraction. It validated him, and made him feel as if he owed it to them to respond. And when they were forthright and open about sex, as Glen Fulwider had been and Marcie proved to be, there was an added lure that defied the old Wesleyan constrictions.

Soon he and Marcie were going out to lunch together, then dinner, then eating in at her apartment, and from the dining alcove it wasn't far to the bedroom. The sex came with no strings attached, and Gould, for a change, was able to take it that way. He and Marcie became a regular thing. Gould was commuting in a company Buick between Syosset and the Bronx, where he and Ruth still lived in her Townsend Avenue apartment. He started telling her he was having to work late, and she believed him because she was used to the long hours he put in when he was working on a project.

There was an outlaw cast to their relationship from the beginning. He and Marcie both knew that an affair between the company's security officer and the one employee without a clearance was a dangerous breach of the rules. Marcie had the power to grant clearances up to the confidential level. Classified papers by the dozens crossed her desk every day. She had the keys to TRG's locked, restricted filing areas. She knew the codes to the alarm systems. But the restrictions were so hard to take seriously in Gould's case that the whole thing seemed like a game.

One thing led to another. They found a concealed spot in the woods at the back of TRG's lot and planted marijuana seeds. The plants grew, and they harvested the leaves, stuffed them in a bag, and Marie took them home to dry. For one summer, they found it one of the more enjoyable vices.

Kennedy's election in November lifted Gould's clearance hopes, and allowed him and Marcie to dismiss the security implications of their relationship.

■■

TRG's laser experiments remained mired in difficulty. The krypton-mercury scheme, employing collisions of the second kind in a gas discharge, had initially looked promising. Experiments had produced some of the necessary population inversion and amplification. But a month later, TRG had put krypton-mercury on the shelf. In its final report of the year to ARPA, the company said it had concluded that the combination would not produce the necessary power, and "because of a limitation in time and funds," it had "shifted virtually all effort to other groups within the project."

In fact, it had been a success, and Daly didn't know it. His team had observed a 1 percent gain; that is, the collisions had produced enough excitation to make the light—the emitted radiation—grow by a single percentage point. That tiny increase was enough to make a working laser. But the information was kept from Gould under his security restrictions, and Daly's team didn't think 1 percent was promising enough. With Gould unable to take charge, what might have been the first successful gas discharge laser died stillborn.

Another potentially successful laser scheme also died at TRG for lack of experimentation. This time, however, Gould was partially responsible. He had suggested that fluorescent dyes could be employed in a laser. The number of resonance lines provided ample opportunity for matching wavelengths with those from a discharge lamp. But when Jacobs proposed that he and Rabinowitz pursue an experiment, Gould said he simply didn't know enough about working with liquids.

Nevertheless, Gould's prescience about potential laser materials continued to be proven in other laboratories. He had suggested that the rare earth elements offered lasing opportunities, and on December 14, 1960, a pair of IBM physicists announced they had demonstrated continuous beam lasers using uranium and rare earth samarium.

At TRG, Rabinowitz and Jacobs continued struggling with optically pumped cesium.

■■

Gould and Marcie tried to be discreet. Inevitably, however, the affair became obvious to the people within Gould's working group, and then outside it among the people with whom Gould was in frequent contact.

This group included Daly, whose animosity to Gould had deepened. Gould's struggles with his security clearance signaled to Daly that Gould was not to be trusted. When he figured out that Gould and Weiss were having an affair, he made sure Larry Goldmuntz heard about it.

The news took Goldmuntz by surprise. It was so outrageous that he at first refused to believe it. Gould had nothing to lose, but he couldn't believe Weiss could be that foolish. Goldmuntz also knew Daly didn't like Gould, and wasn't above trying to sabotage him. But Goldmuntz asked around, and before the year was out had all but confirmed the uncomfortable truth. All sorts of things went through his head. He had visions of his seven-year-old company crumbling. He thought the security boards, if Weiss's affair with Gould came to light, might deem the entire company a security risk. He saw himself hauled before a security board and judged unfit to be trusted. He imagined a revocation of TRG's security clearance, mass layoffs, a restructuring that would force the board of directors to resign, the Air Force killing Project DEFENDER.

He called Weiss into his office. "What's this about you and Gordon Gould?" he said. "The grapevine has it that you're having an affair."

She wouldn't deny it.

"I don't understand how you could do this," Goldmuntz said. "It just won't do. You can have affairs. If you're outside of the security office, you can have all the affairs you want. But this is the one guy who can't get a clearance. You should have had better sense. I hate doing this, but I can't have you in the company."

Gould found Weiss at her desk, packing her vacation photos, holiday cards, and the vase she'd bought to hold the flowers Gould gave her, and knew immediately what had happened. They went to dinner that night, and he promised to help her financially until she found another job. They agreed that, if anything, they were glad they had been found out; now, they could be more open.

Within days, Gould arrived home to find Ruth in a cold fury. She confronted him about the affair with Weiss. Gould confirmed it, and with a grim look and frost in her voice Ruth ordered her husband out of the apartment. Gould left without protest. He felt guilty for having treated Ruth badly, but also relief at having the situation brought to a head.

Ruth's steely dismissal of Gould contrasted with her real emotions. She had invested a great deal in their marriage, and was devastated by what she considered his betrayal. She sought out mutual friends to talk about what had gone wrong, and to seek advice about how to repair things.

Quickly, however, Gould's ouster stretched into what could only be called a separation. He took a ground floor apartment in an Upper East Side Manhattan brownstone, at 329 East Eighty-second Street between First and Second avenues. The location cut half an hour each way from his commute. Not only that, it was a quintessential bachelor pad with a little garden in the back, good for entertaining.

■■

On February 1, 1961, word of yet another working laser made the news. Had it not been for the launching of a monkey into space this one, too, might have made the front page of the *Times*. It was the work of Townes's former graduate student, Iranian-born Ali Javan, working at Bell Labs. The stories said Javan had demonstrated a continuous beam laser using collisions of the second kind, employing helium metastable atoms in a discharge containing a mixture of helium and neon gases.

Again, one of Gould's laser schemes had been put into practice first by someone else.

Javan's laser was a curious case. He dismissed Gould's patent claims and proclaimed, "I am the originator of gas lasers." But he had had little experience with collisions of the second kind in gas discharges at the beginning of his work, when he revealed his inspiration to Townes at almost the same time Gould's TRG proposal landed on Townes's desk. Gould's thinking at that point was probably more advanced. The proposal was apparently the first document to suggest that collisions of the second kind could be used to excite a laser. Javan, meanwhile, had approached another physicist to suggest that they work together.

William R. Bennett, Jr., was teaching at Yale when he received Javan's phone call sometime before Christmas 1958. Bennett had completed his thesis work at Columbia under C. S. Wu, a physicist whose research was so highly regarded that she was mentioned as a candidate for the Nobel Prize. Bennett, for his thesis, had studied the collision transfer of excitation in the noble gases. Javan, who was about to go to work at Bell Labs, mentioned Townes and Schawlow's just-published paper on optical masers. He said, according to Bennett, that he was interested in exploring the use of gas discharges for achieving optical maser action, and wondered if Bennett was interested in collaborating on a paper "in view of your knowledge and experience with problems involving metastable states and transfers of energy."

Bennett felt sure that Javan's information about the possibilities inherent in such transfers had come from sources outside his own research, but from where was unclear. It was clear only that Javan had been reading the professional literature dealing with gas discharges. Nevertheless, he had known Javan since graduate school, liked him, and was intrigued by the prospects for a collaboration.

After the preliminary phone call, Bennett came to New York to meet with Javan. It was during the Christmas break from classes, and Bennett's wife,

Fran, came along. They went to Javan's apartment near Columbia. Javan's girl-friend Marge made it a foursome. Bennett agreed to Javan's suggestion that they collaborate, and at the end of a pleasant evening he left with Javan a pre-liminary copy of his thesis.

Throughout the rest of the winter and spring of 1959 they talked on the phone frequently, and Javan made several weekend visits to Bennett's home in Connecticut to discuss the paper they supposedly were writing together. Dur-ing these discussions, Bennett showed Javan several ways to estimate electron-atom and atom-atom collision probabilities and the lifetimes of excited states. He also gave Javan a summary of his data and calculations on the problem, and told him of his observation, made at Yale, that neon lines dominated the spec-trum in a spark discharge through a high-pressure sample of helium. They didn't only talk physics. At the time, Javan was reading *The Prince,* Machi-avelli's classic about manipulation and the uses of power, and he expressed his enthusiasm for the work to Bennett. One day, Javan called Bennett from Bell Labs.

"He told me that the government was about to classify everything in the laser field because of the Gould proposal," Bennett said. "He said he had been told by the people at Bell Labs to write a paper immediately, that he was being forced by the people at Bell Labs to get something out fast. And he published a paper that July, in *Physical Review Letters,* that had included everything we had talked about."

The paper listed Javan as sole author. Although Javan acknowledged "help-ful discussions" with Bennett and others, including Townes and Schawlow, a good portion of the paper, according to Bennett, contained data based on his calculations. Bennett was shocked and hurt when he saw the paper, which he believed should have carried his name as co-author. In his outrage, he almost decided not to move to Bell Labs, where he had accepted a job. But he had al-ready quit Yale and sold his house, and he had three children to feed. At Bell Labs, he and Javan continued to work together on a collisions laser. Bennett did not know at that point that Bell Labs already had filed a patent application, again in Javan's name alone.

TRG was sweating its contract deadlines and burning midnight oil at exper-iments, while at the Columbia physics department laboratories, more feverish work was being done. It would have been impossible for Bell Labs not to know about the work under way elsewhere. The scientists talked to each other all the time. Nevertheless, Javan professed to feel no pressure to make the experi-ment succeed. He claimed he was aware of no competition to be first. Rather, he occupied a serene area of thought and work. "It was a matter of poetry, of completing a symphony. It only had to be done right," he said.

The experimental work was primarily in the hands of Bennett and an optical physicist named Donald R. Herriott, who remembers Javan's attitude as less cerebral. "One of his big concerns was, 'Hey, other people are on our tails. We've got to get this going,'" Herriott said. "He participated in plans on where we were going, but he didn't do that much to move us ahead."

The laser first worked on a snowy Tuesday, December 13, in 1960. It wasn't publicized at first because Javan and Bennett wanted to submit a joint paper on it to *Physical Review Letters,* whose editors were loath to carry material that had received attention elsewhere, even in the daily newspapers. Bell Labs deferred an announcement until the Physical Society meeting in New York on January 31, 1961. While Javan, Bennett, and Herriott all received roughly equal time at the meeting to talk about their roles in the laser's development, Charles Townes presided at the news conference and focused attention on Javan. He still clung to the rapidly disappearing term "optical maser." But by now there was no question about the device's use in communications. A story in *The New York Times* said modulation of continuous light beams eventually would allow them to carry voice, radio, or television signals.

Javan and Bennett had filed together for a patent through the Bell Labs patent department almost as soon as they had made the laser work. That filing replaced Javan's May 27, 1959, solo application, which Bell Labs had decided to abandon because Gould's patent application had priority. The new application, based on Bennett and Javan's experimental work, described a more limited, primarily pulsed, laser that didn't read on—conflict with—Gould's disclosure. Herriott, who devised the laser's mirrors and a way to align them, was added to the patent retroactively.

Javan left Bell Labs for MIT early in 1962. Bennett left at midyear to return to Yale. Unknown to Bennett, however, Bell Labs had subsequently applied for a patent for a continuous wave helium-neon laser, again under Javan's name alone, and this application did conflict with Gould's. Bennett learned of it only several years later, in the middle of Gould's patent struggles. He was giving a deposition when a patent examiner observed that laboratory notebooks Javan had submitted to defend his claim were not all in the same handwriting. "Whose handwriting is this?" the examiner asked. Bennett found himself looking at his own notebooks.

"I was pretty badly used by Javan during that period. I've never forgiven him," Bennett said of his history of slights at Javan's hands.

Only when the notebooks were revealed as Bennett's did Javan move to amend his patent application to include Bennett as a co-inventor. Later, Herriott's name was added to the still-pending application. By then the helium-neon laser was on its way to ubiquity as the laser used in supermarket checkout lines, and the stories crediting only Javan with its development were years in the past.

21

News of the Bell Labs laser did nothing to improve Gould's disposition. It got worse when he read in the scientific journals that the laser's action had been triggered by the same 1 percent gain that Daly had dismissed in TRG's gas laser as too small. Jacobs and Rabinowitz began to fear approaching him. At first they had talked to him frequently. But as the winter of 1961 turned into spring, and then summer, with no results, he often turned on them impatiently. It did no good to explain that they, too, were frustrated, whipsawed as they were between Gould's descriptions of how an experiment should be conducted and Dick Daly's often contradictory directions.

It wasn't like Gould to take his anger out on colleagues. But he was feeling pressure from all sides. Not only were the experiments going badly, but his efforts to gain a security clearance remained futile. And patent concerns had crept into his thinking, since attorney Robert Keegan had filed papers before the March 22 deadline contesting the validity of Townes and Schawlow's laser patent.

Steve Jacobs, like Gould, was making a thirty-mile reverse commute from Manhattan every day. The increasing pressure to perform often kept him working on experiments deep into the night. "You didn't want to shut the apparatus off, it took so long to get it up to temperature," he said. "So you'd just keep working."

Feeding Jacobs's and Rabinowitz's frustration was the fact that they didn't know why they were failing, only that they were. "We worked very hard to align the mirrors," Rabinowitz recalled. They had to be closely spaced, no more than a few inches apart. Then low-pressure mercury vapor was excited in the vacuum between the two mirrors, and if they were exactly parallel an interferometer—the same Fabry-Perot interferometer that had been Gould's inspiration for the laser—would show a pattern of concentric rings, like a target. Only then could the cesium cell be placed into the device between the mirrors. The slightest nudge could knock the mirrors a tiny bit askew.

"They could get out of line, but we wouldn't know it," Rabinowitz said. "The experiment would simply fail. We were hoping we might see something. But these were 'go, no-go' experiments. You set the thing up, you turned on your helium lamp, and you looked. If you didn't see something, then you had to try something else. We tried pulsing the helium lamp. The lamps were always a problem, because as good as the tube lab was, it was hard to make lamps that were free of other sources of contamination that would make the lines weaker. We would have these horrible setbacks that would blow us away for six months at a time.

"The bad thing, scientifically, was that we weren't learning anything."

So around the middle of the summer, they decided to retrench and try an interim step.

Rabinowitz wanted to see, first of all, if the helium lamp was actually producing amplification in the cesium. They had been trying to produce laser action at the 7-micron transition Gould wanted. It was a stage of excitation where they expected to find a higher population of excited atoms, and therefore more gain, or amplification. But aside from the ongoing problem with the lamps, they had no detectors to tell them if they achieved any gain at all. Three microns was the next higher excitation stage. The gain they could expect at 3 microns was less because they would have fewer excited atoms, but at least they had instruments that could detect it.

Gould resisted. "No, no, just keep working," he said. "Make the flash lamp brighter. I don't want just amplification. I want it to oscillate."

Rabinowitz argued. "We know we have an inversion," he told Gould. "We keep having these terrible catastrophes with flash lamps exploding. Let's switch to something with less gain, where it's easier to work."

Jacobs backed Rabinowitz, and they prevailed. They sketched out a design, and Poulos in the tube lab built a device using two cells. One was tiny, about four inches long, that contained the cesium; the other a much larger cell in which the light would traverse. In the laboratory with the new device, Rabinowitz and Jacobs pumped the cesium cell with the helium lamp, and then looked for light amplification. The transition was at the infrared end of the spectrum, so they couldn't actually see it, but their instruments measured a gain of 4 percent. The lamp had excited the cesium vapor, setting off the chain reaction of emitted radiation that made the light grow that much brighter as it went through the tube, as an ordinary light bulb does when you turn up a rheostat.

This was no history-making event like Maiman's laser, but it was significant. It was the first actual measurement of coherent gain, where someone had taken a light source and determined the amount of amplification. Jacobs and Rabinowitz convinced Gould the three of them should submit a paper to *Physical Review Letters*. The referees at first doubted the measurement results. Calculations indicated the gain should have been higher, and the 4 percent wasn't a huge, knock-your-socks-off gain that erased the possibility of a mistaken reading. They just weren't sure if it was real. But after much arguing on Jacobs's part, *Physical Review* accepted the paper for the issue of December 1, 1961.

■■

That fall, the "secret" classification of Gould's larger patent application was lifted. TRG's contract for Project DEFENDER, however, remained classified. And though it was almost academic now, that fall also saw the beginning of the end of Gould's long effort to be cleared.

Goldmuntz, as well as the lawyers Leventhal and Yarmolinsky, had expected that Kennedy's election would lead to a relaxation of industrial security policies. Nobody wanted spies running around in vital industries, but they hoped, at the very least, for an end to the McCarthy-era exclusions of people like Gould. The new Kennedy administration, however, had produced very little change. Everyone seemed to have forgotten Kennedy's family history. The new president and his family had been close to Joseph McCarthy when the witch-hunting senator was in his heyday. Kennedy, while still a senator himself, had been reluctant to publicly censure McCarthy, and was the last Democrat to do so. And his brother Robert Kennedy, the new attorney general, had served as minority counsel to McCarthy's committee. Anti-communism was embedded in the Kennedy political genes.

What was more, J. Edgar Hoover remained at the helm of the Federal Bureau of Investigation. Hoover was a fervid anti-communist. His policies guided not only the FBI, but filtered down to the agencies that relied on the bureau's opinion about who did, and did not, pass muster.

Six months into the new administration, Harold Leventhal had found the security apparatus as inpenetrable as ever. He expressed his confusion and frustration to Larry Goldmuntz in a June 7, 1961, letter. Its tone was desultory. The lawyer reported to TRG's president that the chief trial counsel for security cases at the defense department "advises that the charging letter will be issued 'shortly,' at the latest within two weeks."

The charging letter was a preliminary step to an arbitrated trial in which the accused—Gould, in this case—would be judged penitent enough to grant a clearance, or too recalcitrant to clear. It was like a religious trial for heresy.

Investigative problems had caused the long delays in resolving the Gould case, Leventhal wrote. He had been told, and passed on to Goldmuntz, that intelligence personnel had been held up by their lack of access to sources and contacts who frequently were traveling out of the country.

Leventhal went on to quote his conversation with the trial counsel, John Davis. "He put it to me: 'The case has been here a long time, and I'm just as anxious as you are to have it disposed of.' He continued: 'It is a very interesting case.' After a pause, I asked: 'Does that suggest you surmise it will be a long trial?' His rejoinder was: 'No. In fact, that's been one of the aspects of the investigation—to condense it so that it will be a relatively narrow issue.'

"Frankly, I don't know what to make of all this so far as timing is concerned," Leventhal continued. "If no charging letter is sent out in two weeks, I'll call on Assistant Secretary Runge." Carl Runge was the assistant secretary of defense in charge of personnel.

More months passed. Yarmolinsky left his practice to join the Kennedy administration as special assistant to Secretary of Defense Robert S. McNamara. That brought his efforts on Gould's case to a halt. Leventhal carried on alone.

■■

The prelude to TRG's breakthrough came when scientists at Bell Labs discovered that curved mirrors could be used instead of flats. This made the mirrors easier to align, so much so that they could now be three feet apart instead of six inches. They still had to be aligned visually, so material that would withstand the corrosive cesium vapor and remain transparent still was necessary. But at least Rabinowitz and Jacobs could now manipulate the ship in the bottle from outside. They had not seen the last of the problems cesium created, though. Barium fluoride would resist the cesium, but it was suitable only for the windows, not for the entire vacuum cell containing the cesium. How, then, to attach the windows, since the cesium would eat any kind of glue with which the windows were attached?

"We built a Rube Goldberg device, and held the windows on with springs," Rabinowitz said. "Daly held us up, because he insisted on using a certain kind of knife-edge seal. Others told us not to use it, but he was a control freak. He was telling us to do it his way, and at the same time saying if we didn't make it work our asses were in a sling. We expected to be fired. But we finally convinced him to do it the other way."

At last they had a device that incorporated a heated cesium cell, a helium lamp with focusing reflectors, and one flat and one curved reflecting mirror.

They could almost taste success. Early in 1962, they moved into a period of intense, exhaustive work. It took all day to bring the experiment to the testing phase. They would begin in the mornings by warming up the apparatus. The cesium had to be heated slowly, the flash lamps had to be working, the water used for cooling had to be safely contained, the mirrors had to be aligned. By the time preparation was over and they were ready to actually run the experiment, it usually was late at night.

The hours extended as the work intensified. Jacobs often caught himself nodding at the wheel on his drive back to the apartment on Washington Square North in Manhattan that he shared with his girlfriend. They decided together that his driving back and forth daily was too dangerous, and moved into a motel in Syosset.

Their regular division of responsibilities placed Jacobs in charge of the optics, including aligning the mirrors, and the mechanical aspects of the apparatus. Rabinowitz dealt with the physics, and the electronics. A technician named Bert Freeman worked with them.

The first American orbited Earth as the scientists drew closer to their goal. Marine Colonel John H. Glenn Jr. lifted off atop an Atlas rocket on February 20, 1962, and went on a five-hour, three-orbit trip around the globe as the nation watched and whispered prayers. Glenn's successful mission galvanized the country.

One Friday night that March, Rabinowitz, Jacobs, and Freeman worked as usual through the evening hours and into the next morning. About two A.M., the device was finally up and running. A detector between the mirror plates was wired to an oscilloscope. Jacobs was using the autocolumnator to align the plates. As he worked, the needle of the oscilloscope clocked along slowly like the tachometer of an idling automobile. Suddenly, out of the corner of his eye, Jacobs saw the needle "go right off the wall."

"Paul, wait," Jacobs said excitedly.

He, Rabinowitz, and Freeman all looked at the detector needle quivering violently at its extreme range. The infrared light was invisible, and they could feel no heat from the microwatt of power, but the reading left no doubt that they had finally triumphed.

The three scientists resisted the urge to drop everything and celebrate. They kept at their work, taking measurements, and with each new test grew more certain they weren't just seeing things, that it wasn't just some oddball noise on the detector. There, finally, was no damping their excitement and exuberance that they had, at last, built the first optically pumped gas laser. Two hours after the first reading, they shut down the machine and went to a local bar to celebrate.

They tried to call Gould, but the phone went unanswered. Apparently, he was away for the weekend.

Rabinowitz and his wife spent a weekend or two every month at her parents' house on Corlear Avenue in the West Bronx. Jacobs was eager to preserve their moment of triumph. He drove to the Bronx the next day to join his experimental colleague. They sat on the back porch in the late winter afternoon and experienced the exhausted afterglow of their accomplishment. Jacobs talked excitedly about how no one else knew what they knew.

"We'll rule the world," he said.

On Monday morning they took their notebooks with their Friday night observations and recorded data to Gould's isolated cubicle in TRG's administration building. Gould looked up from the book he was studying and took in their expressions. He didn't have to be told. "Well, I'll be damned," he said, a smile breaking over his battle-weary face. "You made it work."

Goldmuntz was less restrained when the three of them appeared in his office with the news. He jumped up and danced a jig around his metal desk. And although it was still early in the work day, Jack Kotik ran to his car and headed off to find some champagne with which to celebrate.

22

Gould flew to Washington on May 21, 1962, for a hearing before the Central Industrial Personnel Access Authorization Review Board. His long-delayed case and clearance were largely moot by then; the important early lasers already had been built. Nevertheless, his intuitive genius remained important to TRG's ability to produce laser applications.

He wouldn't have believed it at that point, but the government still had the ability to pull off a few surprises. Gould learned, for the first time, that agents had tapped his phone after he and Glen started attending Joseph Prenski's Marxist study group. Someone in the group had informed on them. An agent who eavesdropped on their conversations testified that he had heard "a lot of sexy talk." Snickers went through the hearing room, and Gould turned red and angry. He turned more angry still when he learned—again for the first time— that the transcript of his appearance before the Special Investigations subcommittee of the New York City Board of Higher Education was in the hands of the review board.

Gould took the witness stand and defended himself as best he could. He repeated yet again the story of his infatuation with Glen, the idealistic beliefs that led him into communism, his gradual awareness of what Soviet communism was about, and his path away from it. But he left the hearing shocked and disillusioned. He had been unprepared for the amount of material the government had in its files about him, and the extent to which he had been branded.

"I had been told that everything that was in my dossier was secret, and that nobody else could ever get hold of such things," Gould said. "So I learned that nothing was sacred in the government, that you can't believe anything they say."

Another ten weeks passed before an agent of the FBI reviewed a brief of Gould's file and the hearing. On August 3, the agent summarized his view that allowing Gould access to work on defense-oriented laser applications "would not be in the national interest."

When Harold Leventhal called Larry Goldmuntz to give him the news, Goldmuntz already had surmised what he would say. "I had the impression that he could never be cleared. Nobody was questioning his loyalty at that point, but at sometime in the past he had lied under oath or refused answers," Goldmuntz said. "The wording was that he had committed an 'indiscretion' that prevented him from being cleared, and there was nothing anyone could do to get him cleared."

"We can take more of your money, but it won't do any good," Leventhal told Goldmuntz, and bowed out of the case.

The legal bills totaled close to $50,000. Explaining the costs of the futile effort to his board of directors, Goldmuntz said, "We paid the doctor, but the baby never got well."

Gould was beyond disillusionment. His only reaction was enriched disgust. He had never named members of the Teachers Union or, earlier, his fellow travelers in the Sixth Avenue Club and the white-collar Professional Club, because he had always assumed that those people were as naive as he was. They were no more dangerous to America's security, and he had refused to ruin their lives to save his own skin. For his trouble, the most creative of the laser scientists was barred from working on his own creation. There was no well-publicized Laboratory Ten to compare with the blacklisted screenwriters and directors who had defied the House Un-American Activities Committee and became known as the Hollywood Ten. But Gould was just as certainly prevented from doing his best work during the height of his powers.

■■

Now Gould turned increasingly to patent matters. He couldn't work on his invention, but it still was his, and he was determined to assert his claims. He saw the laser's value increasing day by day as it found applications that bore out his first predictions. Even one he'd failed to predict he had had a hand in.

More than a year earlier, Gould recalled, he and a colleague had made off with TRG's first laser. It had been highly irregular, given his fight for a security clearance, but Gould had never been one for dwelling on the niceties. And the way things had turned out—the episode had been a prelude to one of the laser's most exciting uses—he thought it had been worth the risk.

The small ruby laser had gathered dust on a laboratory shelf after Dick Daly's group had made it work in the fall of 1960. The laser and the work that went into it remained classified, but it seemed to be entirely forgotten as the company struggled with the other versions.

Late in the year, word reached TRG physicist Gerard Grosof that a group of ophthalmologists at Bellevue Hospital, headed by Milton Zaret, wanted a laser for testing purposes. They were concerned about the laser's potential to cause burns to the eye similar to those caused by viewing a solar eclipse or an atomic fireball, and they wanted to see how it worked on rabbits. Two people were the ideal team for moving the laser, which was small but consisted of two pieces, a power supply and the laser itself.

Grosof says he doesn't remember the details of how the laser got to Bellevue. Gould says Grosof approached him, and he agreed to help. He didn't see how he could lose more to the security authorities than he already had. He was also curious what the doctors would find out. So one night he and Grosof—who had a security clearance—made excuses for working late and around midnight, when the building was empty, they entered the classified area where the laser was stored, carried it out of TRG, and delivered it to Bellevue.

Once the laser was set up, a few days later, the doctors made their first experiment. Nobody knew what would happen. The doctors were nervous about whether the focused beam would reflect into their eyes and blind them, but they didn't wear protective eyegear because nobody knew what would work. Instead, they waited for the countdown that would signal the laser's pulse and at the last second closed their eyes.

Zaret was holding the rabbit. He had to keep one eye open to be sure the rabbit was where it needed to be in the beam's path, and he held it tight. After the countdown reached zero and the laser pulsed, the doctors all gathered around to see what had happened. Zaret eased his hold on the rabbit and it didn't move. It was dead. In their dismay, the doctors started patting themselves to make sure they were all there, and edging toward the exit. Then Zaret realized what had happened—the problem wasn't the laser, it was him. He'd been holding to rabbit so tight he'd suffocated it.

Their fears calmed, the doctors took turns looking at the rabbit's eye through an ophthalmoscope. The neat hole in the retina confirmed the potential for eye damage from a laser beam, and they would write up their findings for the journal *Science*. But as they examined the rabbit they saw something remarkable—the retina is at the back of the eye and there was no damage to the intervening tissue. The doctors took a second turn at the ophthalmoscope. Not a word was spoken as each one in turn considered what he was seeing. They were awakening to one of the laser's most revolutionary uses. Even Gould had not predicted it. Goodwin Breinin was one of the doctors in the laboratory, and he would always remember the thrilling awareness that here was something better than a scalpel.

After preliminary tests, the ophthalmologists had contacted Goldmuntz, the contact with TRG was legitimized, and the laser returned safely to its shelf. The doctors wanted to bring rabbits to the Syosset facility for further optical surgery experiments. Goldmuntz, excited by the prospects, consented. The rabbits, of course, were prolific. They were cuddly little things, too, and people were always taking them out of their cages to pet and fuss over them, and they would get loose and hide. Before long, rabbits were running around the laboratory floor. Finally, Goldmuntz told the Bellevue surgeons, "To hell with this. You keep the rabbits, and we'll bring the lasers to you."

Only a few months later, doctors at Columbia-Presbyterian Hospital used a laser to disintegrate a tumor on a patient's retina, and the era of human optical laser surgery was born.

Other new laser applications were being reported almost every day. Each report added to the growing sense that the laser was indeed a special invention. The stakes in the patent fight were growing.

Gould had, at long last, signed a patent agreement. It assigned any patents deriving from his laser applications filed in April 1959 to the company for the token fee of $1. That was standard patent assignment practice. What was unusual about the agreement was that it split both the costs of prosecuting the patents and any enforcement litigation between the company and Gould. The company's share was 60 percent, and Gould's 40 percent. They agreed to split royalty proceeds along the same lines, and Gould's share of legal costs was to be applied against payments due him in the future.

TRG could well afford what was in effect an interest-free loan to Gould. The company had prospered in the wake of its Project DEFENDER contract as the cold war continued to be good to technology research in the United States. By the spring of 1962 TRG employed almost four hundred people, including a large cohort of physicists, mathematicians, and engineers with graduate degrees.

Gould had always believed that patent royalties were something to fight for. Goldmuntz, too, now realized that the patents that could emerge from Gould's applications represented a potential river of wealth into TRG, and he put the company's energies behind the fight.

23

The claims of inventors occupy a special place in the American saga. Indeed, the promise of invention is embedded in the Constitution, which in Article I, Section 8 assigns to Congress the power "to promote the progress of science and the useful arts by securing for limited times to authors and inventors the exclusive rights to their respective writings and discoveries."

The first patent examiners held other jobs. Thomas Jefferson was the secretary of state in 1790 when he awarded the first patent to Samuel Hopkins of Vermont for a method of making potash to be used in soap. Patents that were issued were published to promote the advancement of technology, while the original patent holders received seventeen-year exclusive rights to their inventions—anyone wanting to make or use an invention during that seventeen-year term had to buy the right to do so from the inventor.

When Gould filed his patent applications, the system of patent issuance was still relatively small. Fewer than a thousand patent examiners occupied cubicles in the north end of the Department of Commerce building near the White House in Washington, D.C. Examiners studied applications in three broad art areas, mechanical, chemical, and electrical, and with the help of a vast library of patents and technical references, issued patents or denied them. Even then, advances in technology were making the task increasingly difficult. Comprehensive patent applications, such as Gould's, frequently embraced more than one and sometimes several inventions, and through a process of dividing the application into parts, continuing some and abandoning others, a set of hybrid patents might emerge.

U.S. patent law is something of a hybrid also. It occasionally allows the Patent and Trademark Office to act as prosecutor, judge, and jury over patents. Recognizing that more than one inventor may conceive similar inventions at more or less the same time, the law sets out within the Patent Office a process to sort out the claims. As the law puts it, "Whenever an application is made for a patent which, in the opinion of the commissioner, would interfere with any pending application or any unexpired patent, an interference may be declared."

The Patent Office may declare an interference independently when an examiner recognizes that two applications conflict. Or either of the inventors may request one. That was what Gould had done in the case of the Townes and Schawlow patent. The requests aren't granted automatically. An examiner must find that the claimed inventions actually conflict—"read on each other," in Patent Office parlance—and declare the parties in interference. Then, assuming both sides can afford it, lawyers get involved. They take depositions, file briefs, and eventually argue their cases before a panel of examiners constituted as a Board of Patent Interferences. And that is only a first step. After that, if one side is still not happy and has some money left, there are the courts. The process of determining who has the rights to an invention can take years.

The slowly ticking legal clock can exhaust an inventor's financial and emotional resources; large, rich corporations and their lawyers have thwarted many an independent inventor's claims until he has lost the will and the ability to fight. But if he has staying power, delays can turn to the individual's advantage. Inventions that come into widespread use while lawyers are fighting can produce vastly larger royalties once a patent issues and the inventor's term of exclusivity begins.

■■

At the Bell Labs patent department, Arthur Torsiglieri had received with surprise the notice that the Patent Office had allowed an interference against the Schawlow-Townes patent. Torsiglieri remained assistant patent director of the Physical Research division; Lucian Canepa, who had written the patent application, had moved to another division, where he was handling logic and memory circuits, and error control systems.

"We had no inkling of it," Torsiglieri said. "We looked at the name, and said, 'Gould? Who is this guy?' "

That was the question Torsiglieri asked Townes when he called him in

Washington at the Institute for Defense Analysis. Townes was nearing the end of his two-year tenure, and still returned to New York almost every weekend to oversee the work his thesis students at Columbia were doing. But his long relationship with Columbia was about to end, for Townes had accepted a physics professorship and the administrative post of provost at the Massachusetts Institute of Technology that would start in the fall of 1961.

Townes also had expressed surprise. He told Torsiglieri he knew Gould. He sketched the history of their acquaintance at Columbia, and his role in reviewing TRG's proposal to the Advanced Research Projects Agency. But Townes said he had no idea Gould had filed a separate patent application.

That fall Townes moved to Cambridge to begin the term at MIT. Then forty-seven and at the height of his career, Townes had reason to believe he was in line to ascend to the presidency of the prestigious institution. Arthur Schawlow, meanwhile, had accepted a professorship in physics at Stanford University, and moved from New Jersey to Palo Alto, California, to start the fall term there.

The second International Conference on Quantum Electronics convened in Berkeley, California. Gould did not attend the conference, and this contributed further to his invisibility.

Over the next months, as Bell Labs' lawyers prepared their case, Townes searched his memory further. He realized that TRG's laser work wasn't widely known since the contract that governed it was classified. He remembered that TRG had insisted on his reviewing its proposal, and his recollections kept coming around to the similarities between the proposal and his and Schawlow's paper. He felt certain Gould had seen their paper, and just as certain that Gould could not have come up with the ideas on his own. Townes actually was beginning to recall Gould as somewhat lazy and misguided. Gould must have cribbed from him and Schawlow. He remembered being disappointed that Gould had failed to credit them, and thought Gould would surely not have based a patent application on the material in the proposal.

He didn't remember quite as well the original work, like Gould's idea to use collisions of the second kind to excite a laser.

Townes also remembered that Gould had talked to him many times at Columbia about patents, and had been very interested in patenting his idea for an optically pumped maser that was contained in the notebook Townes had signed, the idea Townes had later appended to his maser patent. It seemed very much to Townes that Gould was obsessed with patents.

Townes now recalled that every time he and Gould had talked about the laser, he had told Gould what he was working on, while Gould kept his ideas to himself. He and the Bell Labs patent lawyers still were unaware of Gould's extensive notebooks, and in particular the one Gould had written and had notarized in November 1957.

■■

Gould saw his patent claim from the beginning as a fight for the money he knew the laser represented. It was the embodiment, in spades, of his first wish as a would-be inventor—to come up with something useful and make money.

Townes was not motivated by money. Lasers had not yet begun to be manufactured in sufficient numbers to provide much in the way of royalties in any case. Despite its use in eye surgery and a few other early applications, the laser was still a solution looking for a problem. Bell Labs had no intention of licensing the Townes and Schawlow "optical maser" patent anyway. Townes's original maser patent was still in force, and he believed, as did his assignee, the Research Corporation, that with its amendment to include optical pumping it was a general patent underlying all wave amplification devices. Townes's belief was that laser manufacturers, when there were any, would pay to license his maser patent.

Townes's deal with the Research Corporation was that after the attorneys who prosecuted and secured that patent had been paid, he would get 75 percent of the income up to $25,000 a year, or more than triple the $8,000 salary he had made as a professor when the maser was first patented. After that, Townes allowed the percentages to switch, so that he would receive a quarter of the income and three-quarters would go to the Research Corporation to be passed on to universities. He also stipulated that Jim Gordon, who had done most of the experimental work, would receive 5 percent of the maser's royalties for the life of the patent.

For Townes the potential loss was something that to him was more important: his pride and reputation.

Townes's résumé by now stretched to several pages. He was inexhaustible. Between 1960 and 1962, he had served on no fewer than twenty boards and committees. As a member of the Scientific Advisory Board of the Air Force, he sat on the Ad Hoc Committee on Radiation Effects, the Arms Control Committee, the Basic Research Panel, and the Electronics Panel. He was a member of the Visiting Committee of the National Radio Astronomy Observatory. He sat on the Air Force Space Study Committee, the Panel on Strategic Weapons of the President's Science Advisory Committee, the Advisory Committee on Scientific Exchanges of the National Academy of Sciences, and the

Visiting Committee of the National Bureau of Standards. He attended meetings of JASON, a group of academics who monitored scientific issues for the Institute of Defense Analysis. He chaired the Strategic Weapons Panel of the Ad Hoc Group on Optical Masers of the Department of Defense. In addition to all this, he sat on the boards of foundations and mutual funds, and on the Council of the American Physical Society.

At least half a dozen professional societies counted Townes among their members. He belonged to the Physical Society, the National Academy of Sciences, the Physical Society of Japan, the American Academy of Arts and Sciences, the American Philosophical Society, the American Association of Physics Teachers, and the Federation of American Scientists. He was an honorary member of the Optical Society of America, and a fellow of the Institute of Electrical and Electronics Engineers.

Scientific papers issued from Townes's desk with the regularity of automobiles from the assembly lines of General Motors. He also had begun to accumulate a substantial collection of science's most prestigious awards, starting with the Research Corporation's annual award, the Page One Award for Science of the Newspaper Guild of New York, and the Morris Liebmann Memorial Prize of the Institute of Radio Engineers in 1958. The next year had brought him the Comstock Prize of the National Academy of Sciences, the Stuart Ballentine Medal of the Franklin Institute, and the Air Force's Exceptional Service Award. In 1961, he won the Rumford Premium Award of the American Academy of Arts and Sciences, the Arnold O. Beckman Award of the Instrument Society of America, and the David Sarnoff Award in Electronics of the American Institute of Electrical Engineers. He won the Stuart Ballentine Medal again in 1962, the Air Force Association's Science Trophy, and the John J. Carty Medal of the National Academy of Sciences.

These were the marks of a serious scientist. Gould had few of them. It was inconceivable to Townes, or for that matter, anyone in the scientific establishment that Townes personified, that someone who had few scientific laurels could have done anything significant.

So only at its most obvious was the developing battle over patent rights. At its core were the two men, their differing philosophies and needs, and the fault lines they represented dividing invention and science.

■■

Bell Labs' legal team began assembling its defense. The lawyers asked Townes to recount his scientific thinking and his contacts with Gould in the fall of

1957. He produced from his files the September notebook in which he'd drawn the "maser at optical frequencies" as a box. Townes's review of his notebooks revealed his two conversations with Gould that October, when he had asked Gould about intensities of thallium. He recalled conversations in which Gould had agreed that the idea of extending maser action to optical wavelengths was "interesting." He could find nothing to suggest that the student who had been unable to complete his thesis was serious enough to have conceived a working laser. If Gould had thought of the solution to amplifying light waves, why hadn't he said so?

Uncharitable thoughts were not Townes's style. He found himself thinking them of Gould, however. Since Gould hadn't volunteered any information, and kept his invention to himself, Townes imagined that Gould had misdated his notebook to try to slip in ahead of him and Schawlow. He imagined that since the notary who had signed Gould's notebook also was named Gould, there must have been collusion. He imagined anything except the unimaginable truth—that Gould had actually invented the laser.

■■

Motions and procedural actions took up several months. One, which was denied, was Gould's motion to add the larger and more comprehensive of his two patent applications to the interference. Thus the case remained focused on only the smaller, unclassified, application.

Slowly, the outlines of Gould's case emerged to the lawyers at Bell Labs. Keegan's preliminary statement asserted November 13, 1957, as Gould's date of conception and said he had a notebook to prove it. After that, Keegan floated the idea of a settlement. He thought AT&T might agree to license a potential Gould patent; it was the kind of insurance big manufacturers often purchased to head off litigation. But while Gould and TRG were prepared to offer a license inexpensively, they would not do it for free.

Keegan met at a downtown Manhattan restaurant with Hugh Wertz, a lawyer for Western Electric, AT&T's manufacturing and licensing arm. He went over the supporting evidence for Gould, and talked about the crucial notebook. Wertz, however, scoffed at the idea that Western Electric would want a license agreement with Gould. It would be forty years, he said, before Western Electric would have any use for such an esoteric device as the laser. Any Gould patent would expire long before then. Further, Wertz told Keegan, he saw no chance that Gould would win any worthwhile patents in the first place.

Rudy J. Guenther, the head of the Bell Labs patent department, called Torsiglieri to tell him what had happened. There would be no settlement, he said, because "these cases are supposed to be decided on the facts, and we didn't want to appear to be buying him off."

■

Gould, meanwhile, had not given up inventing. He was working at TRG on a number of signal amplifiers, one with a name as quirky as the intuition that conceived it. Gould had been toying with the effects when two frequencies were mixed in a nonlinear medium. Most materials in all regions of the electromagnetic spectrum respond in a linear way when small forces—an electrical field at low levels, say—are applied. They absorb the energy readily, like a rubber band stretching. But when the forces applied reach outer limits, the response becomes nonlinear. The stress can no longer be absorbed, and the materials go haywire. Gould determined that when two frequencies were mixed in such a medium, they would play off each other and the weaker of the two signals would be amplified by picking up energy from the stronger one. He called it the constant loss amplifier, a reference to the energy lost by the stronger signal in the amplification process. The idea looked promising for detecting radar signals, but the name was too confusing. TRG renamed it MARS, for Millimeter Wave Amplification by Resonance Saturation, and won a contract to investigate it from the Rome, New York, Air Development Center of the Air Force. The contract was unclassified.

He also pushed the use of corner reflectors, an idea that had grown out of the struggles Rabinowitz and Jacobs, and Kotik and Newstein on the theoretical side, had had in trying to align flat mirrors. Gould determined that a three-sided corner mirror would act as a retroreflector, and potentially replace one of the end mirrors in the laser. The curved mirrors developed at Bell Labs obviated that idea, but Gould saw other uses. He, Jacobs, and Rabinowitz knocked a hole in the concrete block wall at TRG. They aimed a high-gain laser through the hole at a retroreflector set up in a field half a mile away. The light bounced back perfectly, showing the laser path could be used for secure straight-line communications.

Gould was still seeing Marcie Weiss after Goldmuntz had fired her from TRG. But a new love had appeared on Gould's horizon. She was a tall, incredibly precocious teenager with curly auburn hair, long legs, and a swimmer's strong shoulders. Her name was Tanya Bogart.

They had met early the previous summer. He was sailing on Long Island

Sound when this water sprite appeared from nowhere at the rail of his boat and called out, "Which one of you is Gordon Gould?" Someone in his party pointed to him, and she said, "Here are your socks." She dropped a pair of wet socks on the deck, flipped back into the water, and swam with slicing strokes back to a smaller boat that Gould recognized as belonging to his colleague Jack Kotik, with whom he had been sailing the week before when he had left his socks behind.

"Who was that?" Gould said.

Tanya was seventeen and he was forty.

Soon they were dating steadily, and when she turned eighteen at the end of June she moved in with him on Eighty-second Street. This didn't sit well with Tanya's mother, whose harangues Tanya was constantly fleeing. She set the police on Gould, and dragged her daughter off to live with an aunt in New Orleans. They got together again when Tanya returned to New York three months later, and this time her mother tried to have Gould arrested. Eventually, a sympathetic judge suggested that Tanya's eagerness to move to California, combined with Gould's agreement to pay her tuition at a junior college there, presented a solution, and Tanya headed west.

For all of its harrowing moments, Gould's friendship with Tanya—which would last on and off for many years—rejuvenated him and renewed the creative spark that he was now applying in the laboratory.

24

By April 1963, both sides were taking depositions. Bell Labs was concerned enough to bring in its top patent lawyer, Edwin P. Cave, who had handled the transistor patent. He headed a phalanx of lawyers that included Torsiglieri, Ralph Braunstein, and David P. Kelley.

Keegan had one other lawyer on his side, Donald J. Overocker.

To win the interference, Gould had to show not only that he had conceived the laser first, but that his description of it would have enabled a person "skilled in the art," in the delightfully archaic language that descends from the earliest drafts of patent law, to build one "without undue experimentation." That is, that his description was "enabling." He also had to show that he had pursued his invention with "due diligence," that is, he had not written a description and then set it aside, waiting to resurrect his claim until someone else had filed a similar claim or put a similar invention into practice. He must have been trying either to build a laser or write a patent application.

Ruth was one of Gould's witness. Though they had been separated for more than two years, they were not yet divorced. Indeed, Ruth still entertained the hope that they would be reunited for a second time. She gave her testimony with no apparent bitterness. She spoke of the long hours Gould spent at his desk late into the night in the fall of 1957, the litter of loose-leaf pages upon which he made his calculations, and the inquiries from Townes that impelled him to record his thoughts and calculations in a dated notebook. She said, "He came in in great excitement to say that Dr. Townes had called and had wanted a piece of information and that he was afraid that Dr. Townes might be thinking along the same lines as he was and, therefore, he had better get busy and patent what he had so far."

"What did he call this idea that he referred to?" Keegan asked.

"He called it the laser."

Ruth went on to testify about the trip Gould made to the candy store notary public to have his notebook notarized, and her annoyance that from then on, her husband had neglected his thesis work to concentrate solely on the laser. So great was his obsession, she said, that in the spring of 1958 he had ignored the seasonal rite of readying his sailboat for the water.

She didn't really understand what Gould was working on, she said under cross-examination. Although she worked as a biophysicist in Columbia-Presbyterian Medical Center's radiation laboratory, the physics of her husband's invention were beyond her. Which Torsiglieri emphasized by asking, "What are the Einstein A and B coefficients?"

"I don't know," Ruth answered.

Torsiglieri pressed Jack Gould, the notary public, about whether anything could have been added after he signed the pages. "Well, now, I can't say yes or no," the notary replied. "But as far as I am concerned, those pages were full where I signed on the margins."

"Did you read the contents?" Torsiglieri asked.

"No. I don't have time to read those things."

"Would you have understood it if you tried to read it?"

"I should say not."

Alan Berman testified that Gould had first revealed his ideas for the laser over a bridge game around March 24, 1958, about the time he was leaving Columbia to go to work at TRG. He had gone into more detail at beach parties they both attended with their wives on Fire Island in May, and again in August. Berman, like Ruth, believed that Gould "was working on the laser to the detriment of his Ph.D. thesis."

Maurice Newstein recounted Gould's revelations of the laser to him when the two of them were working in the darkened lab at TRG in May of 1958. Larry Goldmuntz described Gould's refusal to sign a patent agreement and his reluctant revelation that September of the invention he was so anxious to protect. Cave on the Bell Labs side had by then adopted a tone of world weariness with the proceedings. "Just a few brief questions, Mr. Goldmuntz," he said as he began his cross-examination, "and then we will let you get back to something useful."

Gould's testimony consumed three days. Speaking as if he were lecturing one of his City College classes and often offering painstaking detail, Gould described his background and the events leading to his invention of the laser, beginning with his understanding that optical pumping could be used to excite a maser.

"At what time did you complete the necessary steps?" Bob Keegan asked.

"I would say I had it clearly in mind approximately the end of October, or around the first of November, 1957. I was motivated by the realization of how much more important it would be to be able to amplify light than microwaves."

"Did you write down your conception of the laser?"

"I did."

Keegan produced Gould's 1957 notebook. "Is this your first recording of the conception of the laser?"

"It is."

Gould said that his furious work on the laser notebook was triggered by Townes's interest in the intensities of his thallium lamps. "If Professor Townes was indeed thinking about some way of building a laser," he said, "then if I were to establish a date of conception for my idea on the laser—early enough to be of any use to me in patent matters—I would have to write it down as soon as possible."

Keegan held up a copy of the Bell Labs patent. "Are there any similarities between the device that is disclosed in this patent and the device that is disclosed in your notebook?" he asked.

"Many."

"Would you point to any similarities."

"Schawlow and Townes list all the alkali elements and specify potassium in particular."

"Are the alkali elements specified in your notebook?"

"They are."

"With regard to the technique of pumping utilized in the patent, is there any similarity with your disclosure?"

"Identical."

"With respect to the cavity for containing the radiation, is there any similarity between the disclosure of this patent and the disclosure of your notebook?"

"Yes."

The lawyer pointed to Gould's title inscription. "You wrote, 'Some rough calculations on the feasibility of a LASER: Light Amplification by Stimulated Emission of Radiation.' Is that the origin of the term 'laser'?"

"Yes."

As his long questioning continued, Gould pointed out that TRG had managed to build the cesium laser based on his proposal, but that no laser had ever been built using the Townes and Schawlow "teaching," as a patent disclosure is called. Keegan stressed the $300,000 it had cost in a effort to show why Gould had not been able to build a laser on his own. Keegan asked, "When you completed reporting your ideas in this notebook, sometime before November 13, 1957, was it your opinion that you had then devised the easiest and most practical way to build a laser?"

"No. But it was one that I was pretty sure could be made to work."

■■

Braunstein cross-examined Gould. Rather than contest the facts of the case, he tried to paint Gould as greedy, duplicitous, and ethically suspect. He implied that Gould had violated his agreement with Columbia to file a patent report and give the university, and its patent agent the Research Corporation, the rights. He went on to ask Gould whether he had considered taking other jobs after he had gone to work at TRG—Gould said he had, before he and Goldmuntz had worked out their patent agreement allowing Gould to keep a share of the laser. Finally, the lawyer questioned Gould closely about failing to win a security clearance. He wanted to know how many people Gould had approached who refused to testify on his behalf. Gould said Glen had been the only one.

Gould took a sardonic view of Braunstein's line of questioning. When Keegan revisited the question of whether Gould would have left TRG for a higher paying job, he asked, "Would you consider taking a job somewhere else now, if it were sufficiently remunerative?"

"As G. B. Shaw said, we have established what I am, it is only a question of how much," Gould replied.

■■

Townes's deposition took Gould to Boston along with the lawyers, who deposed Townes in his office at MIT. He, Keegan, and Gould were walking to the men's room during a break when Gould said, "Isn't it a shame that we're spending all this time on something that's essentially nonproductive?"

Townes turned to him and said, "Yes, Gordon. Why don't you give up?"

His testimony and Gould's differed on several minor points. Gould, for example, remembered Townes as secretive about his purpose in asking the intensities of Gould's thallium lamps, while Townes said he had been open about working toward an "optical maser."

There were significant differences as well.

Townes placed the fall of 1957 as the time he first talked about the possibility of producing maser action at optical wavelengths. "Sometime in the fall," he
said, was when he had a discussion with Schawlow in which Schawlow "suggested that we should consider a very open structure such as the Fabry-Perot.
And from about that time we worked rather directly and extensively together
on these problems, discussing them together almost every time that I went to
Bell Telephone Laboratories."

This estimate, however, appears optimistic by months. Schawlow kept the
records of his talks with Townes, and when he was deposed later he placed the
genesis of the Fabry-Perot discussions much later, in the late winter or spring
of 1958, after—not before—Gould had written his notebook.

The first firm dates he could confirm for discussions of the "optical maser"
that included the Fabry-Perot cavity structure were April 28 and 29, 1958.
Those were the dates, Schawlow said, on which Ali Javan visited Bell Labs.
The laboratory wanted badly to recruit Javan, and Schawlow had filled out a
special recruiter's report that fixed the dates. Javan later testified that
Schawlow had discussed the laser, substantially as it appeared in the Townes
and Schawlow patent application, during his visit.

Gould was confident in the wake of the depositions. His 1957 notebook was,
in his view, clear evidence of conception, and he knew that he had worked very
hard to convert the information in its pages into a working laser.

The lawyers on both sides reviewed the transcripts of the testimony and
prepared their submissions to the Patent Office's Board of Interferences.

25

A remarkable thing about the laser was that while its abilities were being widely touted, few people knew what it looked like—not the laser itself, but its powerful, pencil-thin beam. Few lasers produced readily visible beams at that early stage, in 1963. Scientists who wanted to dramatize the laser's power were forced to create visual effects.

Steve Jacobs was one of the first to do this. In the wake of the cesium gas laser, he and Rabinowitz had hustled to get their results published. A piece on the laser itself appeared in *Applied Optics*, and they submitted a paper on optical heterodyning using a continuous wave gas laser to *Electronics*. Heterodyning had existed for years in radio. It was a way of combining frequencies, one incoming and one local, so that the incoming signal was amplified and separated so it could be heard. Jacobs and Rabinowitz had applied the same principle to light waves, and demonstrated that a single-frequency, or coherent, light beam could be loaded with information just as radio and later microwaves had been, and that the signal could be separated from the background noise and read. The paper appeared in *Electronics* in July of 1963.

Jacobs had gone to pains to create a dramatic photo for the journal's cover, aiming a laser at a beam splitter where the beam diverged in an X to a reflector, a phase modulator, and a detector where the modulating signal was recovered. The naked eye saw only four separate pieces of equipment sitting at the points of the X with the beam splitter in the middle. Jacobs produced the appearance of red laser beams by darkening the laboratory and blowing cigarette smoke through the beams while he took a time exposure, then turned the lights on for an instant to expose the room and the equipment. It looked like a photographer's long exposure of a busy traffic intersection, in which red automobile taillights were reduced to a continuous stream in four straight, diverging streets.

■■

Laser scientists from around the world converged on Paris that September for the third International Conference on Quantum Electronics. Jacobs, Rabinowitz, and Gould were among the delegation from TRG. Gould's contribution to the conference was a paper entitled "Selective Excitation by Photodissociation of Molecules," an area in which he saw patent possibilities.

Before the conference, military intelligence officials had visited TRG's Syosset offices to lecture the scientists on the information-gathering duties of loyal Americans working on military contracts. They were to be guarded around Russians, who might be KGB agents, and they were to probe the Russians in turn for whatever they could learn. Anything they learned could be reported at a certain room of a hotel in Paris.

Gould was excluded from the meeting. The officials approached him privately, however. They gave him the same briefing, and the same request to find out what he could about the Soviets' work in laser science.

The several hundred conferees met at the headquarters of the United Nations Educational, Scientific and Cultural Organization, on the Place de Fontenoy in a district of government buildings not far from the Eiffel Tower. They sat in a room that reminded Gould of Radio City Music Hall and listened to papers being read, while interpreters gave simultaneous translations in English, French, and Russian.

Gould proved a garrulous and wide-ranging member of the conference who worked hard to find out what the Soviets were up to. One of his contacts was the physicist Aleksandr Prokhorov, who with Nikolai Basov was active in the laser field.

Prokhorov was congenial and easygoing. His colleague Basov, head of the famed Ledbedev Institute, was a committed communist with a corresponding belief in the bureaucracy in which he had risen, but Prokhorov winked at the system. Gould got from Prokhorov the feeling that he shared Gould's dim view of institutions. They talked about things in general in the laser world and, subversively, exchanged congratulations about their ability to use the laser and its potential to win research funds from their respective governments.

Gould also sought information from Basov, but only indirectly. A paper by a member of Basov's research team had appeared in one of the physics journals. It reported a population inversion in mercury—a top-heavy population of excited atoms—that hinted that a very powerful laser was at hand. After the paper appeared, a researcher at General Electric's laboratories had spent a year unsuccessfully trying to duplicate the result Basov had reported. The researcher, James LaTourrette, had then left GE to go work at TRG, and wanted to continue the experiment. Gould knew somebody who was attending a meeting in Moscow before the Paris conference, and asked him to sound out Basov about the paper. The word came back to Gould that the Soviet scientist reacted with embarrassment, and explained that the published figures were based on faulty observation.

"He wouldn't admit in so many words that it was based on error," Gould said, "but the way he reportedly acted told us that was the case. So by asking about that paper, I knew not to reinvestigate that scheme."

Gould, like a loyal citizen with a security clearance, reported his conclusions to the intelligence authorities in their quarters at the Hotel Georges V, and put to rest American fears of a Soviet super-laser.

■

Gould was nearing a crossroads in his relationship with Marcie Weiss. Tanya Bogart was out of convenient reach in California, but other women had appeared on his horizon. One was his new assistant, Marilyn Appel, who had started working as his secretary in the spring of 1962. At first, when they both were dating other people, they had been just friends.

A stronger attraction had begun, as was typical with Gould, on a sailing trip. He had invited Appel and her boyfriend of the moment to go sailing with him and Weiss on Long Island Sound. Appel's yellow bikini first caught Gould's attention, but she also had a wry wit, a clear eye for hokum, and a sharp tongue with which she expressed her opinions unreservedly. And she had an appreciation of Gould's work, having worked at Columbia as a secretary in the physics department before hiring on at TRG. Most of the secretaries went home at the end of the day, but Appel liked to hang out with the physicists and physics students. She was, in her words, "a physics moll."

Gould's relationship with Appel, which would ultimately prove stronger than any of those before, was slow developing. In the meantime, Gould was saying his goodbyes to Weiss. As relationships often tend to do, it ended badly. Eventually, she told Gould she planned to cut her ties to Long Island and move to New Jersey. In time she dropped out of sight.

President Kennedy was assassinated in Dallas that November. The nation paused to mourn, and take stock. The movement for civil rights and an end to Jim Crow laws and enforced racial segregation was building in the South, even as the United States was committing critical levels of troops and matériel to shore up a corrupt and unpopular, but anti-communist, regime in South Vietnam.

By the end of the year, in Washington, the arguments on both sides in the Gould vs. Schawlow and Townes interference were in the hands of the Board of Patent Interferences.

26

The Bell Labs legal team had hinted throughout the depositions that they doubted the authenticity of Gould's 1957 notebook. Keegan's brief to the Patent Office anticipated an attack, saying, "The party Schawlow et al will doubtless infer or imply that the Gould notebook has been altered or may have been altered."

In fact, despite their initial suspicion that it was, as Torsiglieri put it, "some sort of fabrication," Bell Labs' lawyers had concluded that the notebook was real. They attacked nevertheless, but only indirectly. Braunstein's brief argued, "If there was derivation in the case, it seems far more likely that Mr. Gould, a long-time graduate student, derived the invention from his conversations with Dr. Townes, the internationally recognized scientist. . . ."

Schawlow rarely received such plaudits from the Bell Labs lawyers in their filings. Keegan, for that very reason, liked the fact that Schawlow's name came first. He enjoyed referring in his legal papers to "Schawlow et al," leaving Townes's name off entirely, because he suspected that it peeved Townes to play second fiddle.

If Gould's notebook was real, Bell Labs had to attack on another front. The next question was adequate disclosure—whether someone "skilled in the art" could have used the notebook to build a laser. The only people skilled in the art of lasers at the time Gould wrote his notebook were highly skilled indeed. They were physicists with not only a thorough knowledge of quantum theory and its mathematical underpinnings, but also with experimental skills developed through microwave spectroscopy, a knowledge of optical physics and instruments, a vast familiarity with the energy levels of the elements, and a corresponding skill in engineering that would allow them to assemble the necessary components and defeat the chemical reactions of the lasing elements. It is unlikely that more than a dozen people in the country, and not many more than that around the world, were skilled enough to build what Gould was writing about when he first put pen to paper.

Bell Labs' expert witness, predictably enough, said a laser could not have been built based on Gould's notebook. Not surprisingly, he was a Bell Labs employee, Dr. C. G. B. Garrett. Gould's experts were arguably more objective, and certainly more confident. Peter Franken, Gould's former laboratory partner, who was now at the University of Michigan, and Dr. Peter Bender of the National Bureau of Standards, each said he could have built a laser from the information Gould provided.

But the Bell Labs brief conceded, "The issue of first conception remains a complicated one."

That left diligence. Here Bell Labs leveled its heaviest attack, arguing that Gould's evidence was "vague, indefinite, and uncorroborated." Ruth's testimony that Gould was working four and five nights a week on the laser was unreliable, Braunstein wrote, because Gould only told her he was working on the laser, and she didn't have the expertise to evaluate it accurately. Even those who could comprehend his ideas had only Gould's word that he was working on them. "The credibility of Gould's own testimony and that of his principal witnesses is questionable," Bell Labs argued.

Gould's habits worked to Bell Labs' advantage. His pattern was to think and study for a long time before he wrote anything, and then record his thoughts and findings in bursts of furious activity. Consequently, there were gaps in his notebooks when he seemed to be doing nothing. Townes's notebooks, by contrast, were practically a daily diary of thoughts, contacts, and ideas, a kind of scientific stream of consciousness.

The most important period in which Gould had to show diligence was the period starting just before Schawlow and Townes's July 1958 filing date, to his own filing date the following April, a period of over eight months. Gould had testified that he went to work at TRG because he thought he could build a laser there, evidence in itself that he was diligent. Bell Labs countered by arguing that he had not told Goldmuntz about the laser until six months after he had joined the payroll.

"And why was Exhibit 9," Braunstein wrote, referring to the notebook that

Gould compiled as he prepared TRG's laser proposal to ARPA, "witnessed for the first time on August 28, 1958, if Gould actually spent his free time from March to June working on it?"

Bell Labs also attacked Gould's tendency toward wariness and secrecy. Keegan had argued that "neophyte inventors are notoriously rather suspicious." Bell Labs waxed incredulous that Gould could have been "suspicious of Dr. Townes and hence reluctant to discuss his laser ideas with him," when Townes had freely told Gould about his ideas for an optically pumped maser. (No arguments were made on either side about Townes's amendment to his maser patent to include optical pumping.) The implication of the Bell Labs' brief was that Gould didn't have any laser ideas. This conveniently overlooked Gould's understanding of the laser's many uses, which was being borne out by events.

Bell Labs consistently chose language that was harsh and dismissive, mocking in its tone, and entirely speculative. The gap of three months in signatures witnessing Gould's second notebook, Braunstein wrote, was just what you'd expect. Procrastination "would be a natural enough failing for a man who had been a graduate student for at least nine years and had spent six of those years working on a doctoral dissertation he never finished."

He used Alan Berman's testimony to harp more on Gould's failure to complete his Ph.D. requirements. It would be natural for Gould to have lied to Berman about what he was working on, Braunstein suggested. "Too many men dissemble when asked why they left graduate school after many years without obtaining the long sought Ph.D. This is especially true where the questioner is a former student who obtained his degree and began a successful career, and the one questioned is a perennial graduate student whose progress toward the degree was slow and uncertain."

Had Gould obtained his degree, Bell Labs' brief would have been reduced by several pages.

Braunstein argued that Gould produced more evidence of what he did not do than what he did during the time in which he needed to show diligence. Ruth's testimony, and that of Geoffrey and June Gould, that Gordon "did not work on his sailboat during the spring of 1958 is grossly irrelevant," Braunstein wrote. "Failure to work on a sailboat, as well as giving up a weekly bridge game, are consistent with a great many other courses of action," but "there is nothing to show what he was doing besides his own testimony." Braunstein treated the absences from TRG, during which Gould had said he was working on the laser, the same way, as showing not what Gould did, but what he didn't do.

"Had he intended to leave no trace of his work, Mr. Gould could hardly have been more successful in concealing it," Braunstein wrote sarcastically.

"Few men who *try* ever succeed in hiding their affairs as completely as Gould appears to have done in the critical period," he concluded.

Keegan argued that Gould had been more than diligent. He had worked on the laser single-mindedly for the eighteen months between conceiving the laser and filing his patent application, making personal and academic sacrifices in the process. He stressed the difficulties facing an inventor in Gould's shoes. Gould's, he wrote, was the diligence of "an individual inventor of very limited resources."

And he hinted at the battles faced by every small inventor. If Gould's evidence could not convince the Patent Office, Keegan wrote, "then it must be concluded that no individual unsophisticated inventor of limited resources can ever hope to prevail in a contest of invention priority against a large corporation and its inventors."

■

The Board of Patent Interferences, composed of three examiners drawn from a large pool within the Patent Office, read the testimony and briefs, heard supporting arguments, and in March of 1964 ruled in favor of Bell Labs, upholding the patent award to Schawlow and Townes.

Keegan wasn't that surprised. Showing diligence on Gould's part had been a difficult challenge. But he was taken aback by the focus of the board's ruling on conception. There, the examiners had seized on Bell Labs' argument that Gould had not specifically stated that the side walls of the cavity were nonreflective and, in fact, transparent, and that therefore, someone of ordinary skill in the art would not have been able to construct a working laser.

"I didn't state it in so many words," Gould had testified in his deposition, "but it is implied that light from the external discharge must be able to get into the laser tube to excite the atoms therein."

"Isn't that obvious?" Keegan had asked.

"Yes. And in describing the effects of the plane parallel mirrors in selecting for amplification those waves which are plane and parallel to them, I discussed the diffraction losses for waves other than those which are plane and parallel, and such losses wouldn't take place if they didn't escape from the side members."

Keegan had not belabored the point with Gould, and Braunstein had not addressed it at all in cross-examination. Peter Franken, however, had testified at length about his understanding of the laser tube's transparent side walls. Bell Labs had attacked Franken's assumption that Gould was describing an open Fabry-Perot cavity when he hadn't actually said so, and sneered that Franken's reasoning was "fallacious and circuitious."

"It was a phony argument," Keegan said. "Bell Labs' lawyers and their expert witnesses knew that notebook had clearly described a laser medium in an optical resonator having only two reflective areas at the ends of a gas tube. But the expert was careful to use lots of scientific jargon to minimize the board's understanding of the Gould disclosure, and the examiners were incapable of understanding the nuances of the testimony."

■■

Gould saw the ruling as a temporary setback, and Keegan filed notice that the inventor and TRG were appealing the board's decision. The next step was out of the Patent Office to the courts. Keegan could have chosen either the U.S. District Court for the District of Columbia, or the U.S. Court of Customs and Patent Appeals, designed to hear the specialized cases customs and patent law created. That was the one he chose. But by the spring of 1964 Gould was fighting not one, but four, interference cases.

The accelerated pace of laser development was bringing new laser claims to the Patent Office almost every day. So comprehensive was Gould's application that the office had begun reviewing other laser-related applications against his. Examiners saw conflicting claims in three, and had exercised the Patent Office's prerogative of launching interferences.

In each of them, Gould was again pitted against large corporations with virtually unlimited resources. Two involved familiar opponents—Bell Labs, and the laboratory's aggressive team of lawyers. Javan vs. Gould would sort out the competing claims to the continuous wave gas discharge laser on which Ali Javan, ignoring his collaborators, had filed a solo application. Two other Bell Labs scientists, Arthur G. Fox and Lynden U. Kibler, had filed an application for a patent on the use of Brewster's angle windows to reduce reflection losses in a laser, and the Patent Office had initiated an interference in this case in 1963. The third of the Patent Office–initiated interferences set Gould against Robert W. Hellwarth, a scientist at Hughes Research, over the Q-switch, the scheme that Gould had identified in his patent application for allowing the energy to be built up in a laser before being released in a giant pulse. Hughes had filed the Hellwarth application on August 1, 1961, more than two years after Gould's was submitted to the Patent Office.

Gordon Gould at Union College, where he learned to love the study of light and the beauty of optical phenomena.
(*courtesy Gordon Gould*)

Gould lectured in physics as a senior, filling in when professors were absent.
(*courtesy Gordon Gould*)

The Pupin Physics Building at Columbia University, where Gould studied after World War II.
(Columbia University Archives and Columbiana Library)

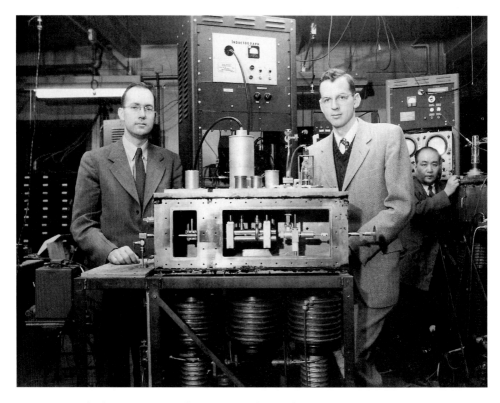

Charles H. Townes and James P. Gordon, with the maser at Columbia.
(Columbia University Archives and Columbiana Library)

Gould's first laser notebook, finished November 13, 1957, in which he coined the word *laser* and set out its essential elements.
(courtesy Gordon Gould)

Arthur L. Schawlow, who received the first laser patent with Charles Townes, his brother-in-law, in 1960.
(courtesy Arthur L. Schawlow)

Donald R. Herriott, Ali Javan, and William R. Bennett, Jr., with the gas discharge laser they developed at Bell Laboratories. *(courtesy William R. Bennett, Jr.)*

Lawrence A. Goldmuntz, the founder of TRG–Technical Research Group–where Gould proposed wide-ranging laser ideas. *(courtesy Barbara Goldmuntz)*

TRG partner and theoretician Jack Kotik holds the plans and Goldmuntz the shovel at a mid-1960s groundbreaking ceremony for TRG's new building and laboratory. *(courtesy Jack Kotik)*

Physicist Steve Jacobs, with the optically pumped cesium laser that he and Paul Rabinowitz built at TRG from Gould's proposal to the Defense Department.
(*courtesy Steve Jacobs*)

The optically pumped cesium laser first worked in March 1962.
(*courtesy Steve Jacobs*)

Gould with colleague Ben Senitzky at TRG, with the MARS (Millimeter Wave Amplification by Resonance Saturation) amplifier he invented. Gould called it the "constant loss amplifier."
(*courtesy Gordon Gould*)

Gould and Marilyn Appel aboard Gould's boat, *Wonny la Rue,* in 1967.
(*courtesy Gordon Gould*)

The picture on the monitor is carried over a CO_2 laser beam, a system Gould developed at Optelecom in 1977.
(*courtesy Gordon Gould*)

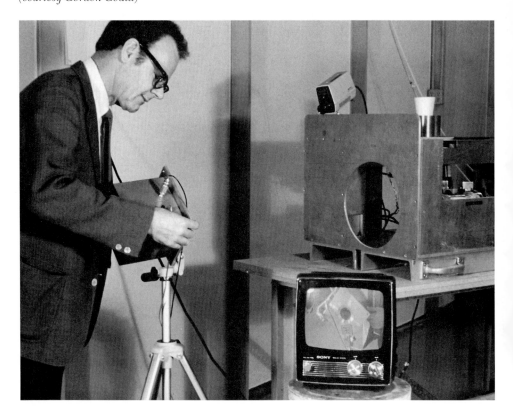

Laser light carried over an optical fiber; Gould in the laboratory at Optelecom.
(© *John Neubauer*)

Eugene M. Lang, the founder of Refac, the technology development firm that licensed Gould's patents. *(courtesy Eugene M. Lang)*

Kenneth G. Langone put together the financing that allowed Gould to pursue his long patent fight. *(courtesy Kenneth G. Langone, © Bachrach)*

Patent attorney Richard I. Samuel and his partners spearheaded the long series of litigations. He bragged that "we invented the laser patent application." *(courtesy Richard I. Samuel)*

Physicist Paul Rabinowitz worked with Gould at TRG and the Brooklyn Polytechnic Institute. Here he's pictured with Gould and Appel. *(courtesy Gordon Gould)*

Gould in 1986, the year before he won his crucial laser patent victories. *(© Tim Wright)*

27

As that spring ran into summer, Gould moved to bring closure to his long-running separation from Ruth. They had agreed informally to an alimony arrangement under which Ruth, who remained a well-paid assistant professor of radiology at Columbia, would receive no alimony as long as her salary remained higher than his. But if he started at last to make more than she did, he would pay. The amount remained unspecified, but Gould anticipated that one day soon he would prevail in his patent fights and begin receiving royalties from the growing laser industry. He thought it would be a good idea to be divorced before that happened.

Divorce, however, was next to illegal in New York State. The Roman Catholic–dominated State Assembly had set adultery as the only grounds. Couples who agreed they wanted to divorce couldn't just make up a confession, because adultery also was against the law, and besides that, judges looked unkindly upon perjury. That limited divorce to the relatively wealthy, who could live in a Nevada hotel for six weeks in order to establish residency, or make arrangements with a lawyer to fly to Mexico for an overnight decree.

Ruth saw the state's strictures on divorce as evidence that she and Gordon still might work out the differences between them. Despite their alimony discussions, she held out an almost desperate hope that they might reconcile.

Gould, however, was happy being unencumbered. Ten and twelve friends at a time gathered for lively dinner parties at his Eighty-second Street apartment. Laughter, conversation, and the clink of ice in glasses spilled out into the garden. Gould had found a Vietnamese-born French chef who sometimes cooked for his parties. She had trained at the Paris Cordon Bleu, and her dishes were exquisite. She could bone a squab, put it back together, and bake it with herbs to produce flavors that brought gasps of wonder from his guests.

The chef was alone in New York with a small child. Her husband had gone to work in Washington, leaving her to support herself at her job in a hotel restaurant kitchen. Soon after she started cooking for Gould, she invited him to her home for dinner. In time they began a quiet, gentle, and occasional affair that occurred around these dinners, and then simply ended—until Gould opened the door one day a year later and found himself facing her angry husband. They argued, and it emerged to Gould's embarrassment that the chef had attributed to him her sexual awakening. "You ruined her," the husband said.

Though he was maintaining the appearances of a bachelor life, and dating several women, he still wanted a clean break from Ruth before he started making money from the laser.

Even a Mexican divorce required an agreement of both parties, and a valid separation agreement that had been in force for at least a year. Only then would New York recognize, under full faith and credit, a divorce granted in another jurisdiction.

When Gould told Ruth he intended to get a divorce in Mexico, she was devastated. She called friends with whom she and Gould had played bridge when they were together, and begged to talk. The sympathetic couple drove to the Bronx to pick her up on a balmy Sunday in September, and over the course of several hours at their home overlooking the Hudson River she cried out her grievances about Gould and their marriage. Even after almost four years of separation, she was crushed by his rejection and the breakup of the relationship.

Gould pursued the matter through a New York lawyer. In time he flew to El Paso, Texas, where he stayed overnight in a hotel. The next morning, he climbed onto a shuttle bus with other would-be divorcées and rode across the border to the Juarez courthouse. When he got off the bus he was glad he wasn't a woman; they encountered a gauntlet of jeering men making kissing noises and calling obscenities. His Mexican lawyer picked Gould out of the crowd by the blue documents folder he was carrying. They went before a judge who, Gould realized from the man's utter boredom, did nothing all day long but sign divorce papers. The judge signed Gould's papers without reading them. Gould paid the lawyer and returned to New York unmarried once again.

But by summer's end, despite the official seal upon his freedom, Gould's relationship with Marilyn Appel was deepening. They first made love during a cruise in his sailboat on Long Island Sound, when they took shelter from a storm in a harbor at Block Island. But, for a change, they had other mutual interests, too. They both loved music; Marilyn, like Gould, was a choral singer. They loved Manhattan, with its stirring variety of music in all its forms—choral and orchestral concerts, the thrilling religious oratorios of Bach and Handel, even the jangle of bands at Greek night clubs where Gould was prone to misbehave on ouzo. Marilyn had left TRG and was attending law school at New York University while she lived on Sixteenth Street just west of Seventh Avenue, not far from where Gould had lived with his first wife. Though she spent a lot of time with Gould at his Eighty-second Street apartment, they didn't think at first about moving in together. Their relationship was still relatively new, and Gould's history told him to resist that kind of permanence. Marilyn, in any case, was making no demands on him.

■■

The Nobel Prize rumor mill begins to stir in early October of each year. It is especially active in the world of physics, as well as in chemistry, biology, and medicine, the other scientific worlds where major researchers know one another and follow one another's work. The same researchers, as well as previous Nobel winners, nominate the candidates among whom the Royal Swedish Academy of Sciences will choose. Leaks occur. The Swedish press, and then the world's, starts trying to develop background material on potential winners for the profiles that will run when the announcements come.

Arthur Schawlow was teaching a freshman physics class at Stanford when a photographer edged into the room. Then a woman from the university's public relations office interrupted him with the whispered news that he was a rumored winner of the 1964 Nobel in physics, along with Townes and Maiman. Schawlow turned to the class and ad-libbed an excuse for the photographer's visit. The Nobel announcements were to be made the next day, and that night Schawlow slept poorly.

In his work with Townes, Schawlow could reasonably expect to be a Nobel Prize contender. Bespectacled, heavy-set, his round face often expressing quizzical amusement, he was somewhat less ambitious than many of the others. He remembered both Basov and Prokhorov lobbying for the honor in Paris the year before at the international quantum electronics conference. Basov had approached him and said, "Prokhorov and I want to talk with you and Townes about who should get the Nobel in this field," as if the prize were the last shrimp on an hors d'oeuvre tray.

While Townes was adding administration to his lengthy résumé as the provost at MIT, Schawlow was emerging as a teacher who could demonstrate the laser in popular terms. Two years earlier, he had appeared on a San Francisco television show called "Science in Action" and used a laser beam to break a balloon rigged up with fins to resemble a rocket ship. Then, on a trip to the San Francisco Zoo with his autistic son, he saw children carrying blue balloons inflated inside clear ones, and realized that a laser beam would break the blue balloon without harming the clear one. This demonstrated that the color of the laser light, and the light-absorbing properties of the target, were important to laser applications. Later, he fitted a child's toy "ray gun" with a portable ruby laser, and used the laser gun in demonstrations for his students, breaking balloons within balloons in an analogy to laser eye surgery.

Schawlow also claimed to have invented the "daser." This device, he jokingly contended, would allow lighthouses to be used in daylight by amplifying darkness.

As far as practical applications of the laser went, Schawlow believed low-power lasers would make good typewriter erasers, and had applied for a patent for that use.

But Schawlow's good humor was not to be rewarded the next morning, when the Nobel prizes were announced. The news from the Nobel Prize Committee breaks at about four A.M. on the West Coast of the United States. Unable to sleep, Schawlow was turning his radio dial when he heard the prize had gone to Townes, Basov, and Prokhorov.

The Stanford public relations department told Schawlow later it was very sorry.

Townes was also on the West Coast when the announcement came. He had been named, earlier in the year, chair of NASA's Science and Technology Advisory Committee for Manned Space Flight, and he was at Cal Tech in Pasadena heading a meeting on the Apollo program. The phone rang at Townes's house in Cambridge at seven in the morning. It was a representative of the Nobel committee, wanting to speak to Townes. Townes's wife refused to tell him where her husband was, but relented when the man explained why he was calling. But when the caller mentioned the names of Basov and Prokhorov, Frances Townes blurted, "Oh, no, not the Russians!"

The media started calling, and Mrs. Townes kept saying, "Oh, no, not the Russians!" She told reporters the Russian scientists had never originated anything. She was not far off the mark; Basov and Prokhorov had not published as comprehensively as Townes. They had reached in a crude experiment a vestige of his maser work, but this did not mollify Frances. By the time reporters began calling Townes to line up interviews, they were asking him why his wife had been so upset that the Russians shared the prize.

Townes fell back on vague diplomacy. "Well, half the prize is not as good as the whole prize," he said.

The controversy was indeed diminished by the way the award was split, half to Townes, while Basov and Prokhorov shared the other half. And Townes, who was as familiar as anyone with the work of the two Russians, described them as "fine scientists" who "quite properly share in the award." The language of the Royal Swedish Academy in naming the three scientists—the most who can share an award under Nobel Prize rules—stressed their "fundamental work in the field of quantum electronics, which has led to the construction of oscillators and amplifiers based on the maser-laser principle."

The Royal Academy didn't quite say the award was for the laser, but people assumed it anyway. That's what Gould thought, in any case, and it annoyed him. He didn't begrudge Townes the Nobel; from Gould's point of view it was just another star—although obviously the centerpiece—in a crown that was already royally bejeweled. Gould's problem was that he was still fighting to prove that he had invented the laser, and the Nobel tended to validate Townes's position while the issue remained, to Gould's mind, undecided.

For his part, Townes did little to allay the assumption that the laser, not its scientific underpinnings in the more limited maser, had won him the Nobel.

The title of his Nobel Prize address, delivered in Stockholm on December 11, was "Production of Coherent Radiation by Atoms and Molecules." The reference to molecules was a subtle attempt to justify his, and Schawlow's, persistence in referring to "optical masers." If only they could convince the scientific world and media that the "m" in maser referred to molecule, not microwave, then an optical maser made sense after all. But it was far too late to erase "laser" from the public vocabulary.

28

Townes's share of the Nobel for physics was about $27,000, a healthy amount of money for a research scientist in 1964. Gould, too, finally was about to realize his dream of making some money from the laser, although in a far different way.

TRG had grown beyond Larry Goldmuntz's wildest dreams. It had continued to land new research projects. At the same time it branched out into a small side business manufacturing solid-state lasers of ruby and neodymium YAG, the latter an artificial crystal combining yttrium, aluminum, and garnet, doped with neodymium in tiny quantities, which provides the active lasing element, as does chromium in ruby. These lasers were used as tools for working metals and other materials. The company's ALWAC computer hummed away at the design of high-performance radio antennas and ship hulls with minimum wave resistance. In 1963, its sales totaled $6 million and its net profit almost $200,000. TRG's laser sales, along with its expertise in high-technology research, its talent roster and patent portfolio, and its government contracts, made it attractive to other companies.

Control Data Corporation of Minneapolis, Minnesota, saw TRG as a likely merger candidate. The computer company's executives believed the laser was the tool of the future for making computer chips and other parts. Bill Keye, who headed Control Data's government systems operations, saw TRG's research contracts as a chance to "add a third leg" to its government systems groups in Minneapolis and Ottawa, Ontario, Canada. Control Data president William C. Norris sent out feelers, and Goldmuntz said he was ready to listen.

As talks proceeded during 1964, Goldmuntz had called Gould and suggested they discuss Gould's patent agreement. When they met, Goldmuntz pointed out that while TRG was making money in the laser business Gould, who was responsible, was realizing no benefit.

"If we're going to sell lasers, you should get a piece of that," Goldmuntz said.

When Goldmuntz suggested renegotiating their patent agreement, Gould's natural suspicion of authority bubbled up. He listened warily as Goldmuntz continued. "What you have now doesn't protect you," TRG's president said. "You've got no way to enforce it if somebody turns out to be hostile. Your patent is still in application, and if you're ever going to make any royalties, they're a long way down the road."

Gould's leftward leanings had moderated, but he still innately distrusted corporations and bureaucracies. Executives weren't supposed to act the way Goldmuntz was acting. He put off finalizing a new deal, but after weeks of insistence, as his interferences dragged on and the deal with Control Data neared reality, he signed a new agreement. It would pay Gould 1 percent of the gross from all lasers sold by TRG, regardless of whether or not he ever obtained a laser patent. It also required TRG, and Control Data if the deal went through, to pursue Gould's patents. Most importantly, it obligated any future buyer of the laser business to pursue the patents.

Control Data announced its plan to acquire TRG on September 24, 1964. The sale was finalized during the fall, and was announced four days before Christmas.

The merger to which Goldmuntz and his board of directors agreed took the form of a stock swap, 205,000 shares of Control Data common stock for the assets and business of TRG. Gould's stock options, his reward from Goldmuntz when TRG won the million-dollar ARPA contract, exchanged for Control Data stock, promised to lift him from relative poverty to paper affluence for the first time since he had decided to make a career as an inventor. He would collect perhaps $10,000 a year in laser royalties under his new patent deal. That was a trickle of money by comparison with the Control Data stock. Gould's new net worth allowed him to pursue in grand fashion one of his favorite recreations. He went shopping for a boat, and soon was the proud owner of a forty-five-foot Pearson ketch. The name he had stenciled on the stern was that of his mother's pirate ancestor, corrupted by tidewater Chesapeake pronunciation: *Wonny la Rue.*

■■

The Court of Customs and Patent Appeals scheduled Gould's appeal in his interference against Schawlow and Townes for December 8. News of the impending hearing placed Gould's name, and his claim to the laser, in the public eye for the first time. *Business Week* devoted several pages to it at the end of November. The magazine made the most of Gould's story. It dubbed the conflict the "candy store patent case," called Gould a "controversial scientist" with "a stormy domestic life," and touted the laser as "one of the most brilliant research developments of the century." To *Business Week,* Gould's claim recalled the beginning of Dr. Lee De Forest's twenty-year fight for the rights to the three-element vacuum tube, the cornerstone of early radio and television.

Trade magazines including *Electronics* and *Product Engineering* also weighed in with stories in advance of the hearing. The latter summed up the stakes in a single sentence, when it said the case "touches every raw nerve in science and technology: patent ownership, Federal security, academic-industrial conflict of interest, research funding, and the rewards of fame—and perhaps fortune."

The Washington Post and London *Daily Telegraph* were among the papers that offered abbreviated versions of Gould's juicy tale in the days before the court's panel of five judges heard arguments. A *Post* columnist, David Fouquet, called the case "a strange and tangled scientific paternity suit . . . over who fathered what has been called one of the most revolutionary discoveries of the century."

The stakes were increasingly high, for by then the laser was becoming the all-purpose tool that Gould had predicted, employed in manufacturing, communications, surgery, radar, photography, computer technology, spectroscopy, and other fields.

Bob Keegan had outlined his arguments in his appeal brief filed in September. He said the Board of Patent Interferences erred in several areas. The examiners were wrong in saying Gould's 1957 notebook did not disclose an operative embodiment of the laser because he hadn't specified that he intended the side walls to be transparent. It was obvious that an optically pumped laser would have transparent side walls. The examiners also erred by ignoring the testimony of Gould's experts that one skilled in the art would have understood how to build a laser from the notebook.

Keegan also attacked the board's findings on diligence, saying the examiners should have credited Gould's testimony and also that of his ex-wife and his experts who had testified that he had disclosed the laser to them. Further, he said Gould should have gotten credit for diligence by going to work to try to reduce his idea to practice, rather than first attempting to file a patent application.

He went on to argue that the board had imposed an unreasonably high standard of diligence, not the "reasonable diligence" required under patent law, and that Gould had "applied all his available resources" to reducing his invention to practice. The difficulty he faced was emphasized, Keegan said, by the fact that Bell Labs, "probably the largest research and development facility in the world, was not able to build a laser until long after Gould's filing date of April 1959," and even then, not before Maiman did it at Hughes.

"In many respects the circumstances surrounding the Gould invention are unusual," Keegan wrote in a case of exquisite understatement. "It is unusual that an independent inventor in these days of corporation research laboratories would be the first inventor of this very significant device. That it would be his first invention of any note is also very unusual. His inexperience in patent matters led him to corroborate his conception of invention in an unorthodox (but effective) manner," he continued, referring to the candy store notarization of the notebook. "His undertaking a project to reduce to practice the laser on his own time is similarly quite unusual, as was his success in his initial objective of finding laser devices which could readily be reduced to practice.

"None of these unusual aspects of the circumstances of Gould's invention detract, however, from his claim to priority of invention. By the very nature of the process of invention the rules for determining priority must allow for unusual circumstances and not be restricted to a pro forma invention process such as would be expected to take place in a corporation research facility."

The court's panel of five judges convened in Washington at the beginning of the second week of December to hear oral arguments. They heard points both sides had made before. Keegan argued that Ruth Gould's testimony proved diligence, and said that Schawlow and Townes hadn't proved they had come up with the crucial Fabry-Perot structure until April of 1958, more than six months after Gould had drawn it in his notebook.

Arthur Torsiglieri argued the case for Bell Labs. He said Ruth had only Gould's word that he was working on the laser, since she couldn't have understood it, and implied that he purposely blurred the distinction between the laser and the light-excited maser.

"You don't think he lied to her, do you?" asked Judge I. Jack Martin, one of the five judges.

"Well, we don't come out baldly and say that, no," Torsiglieri backtracked. "There is no evidence he told her he was working on a Fabry-Perot laser."

If Gould really had been diligently trying to build a laser, he continued, he would have been trying to raise money. Torsiglieri said Hughes had spent

$100,000 making the first working model of the laser. "It takes $20,000 in equipment just to tell you if you've got it working," he said.

"That puts the so-called attic inventor at a disadvantage, doesn't it?" asked the panel's chief judge, Eugene Worley.

Keegan answered Torsiglieri by rejoining that much of Gould's conceptual work had been aimed at determining the least expensive approach before revealing his concept to Goldmuntz at TRG.

Torsiglieri picked at Gould's visit to Darby and Darby in January of 1958, when Gould made general inquiries about his patent rights versus Columbia's.

"If the patent had been filed then, there would be no question that he would win," Torsiglieri said.

"But is that a reflection on him or his counsel?" asked Judge Worley.

Neither Keegan nor Torsiglieri could tell from the judges' demeanors which way they were leaning. The panel took the case under advisement, and Gould once again found himself waiting for a ruling that would decide his future.

29

Dramatic and unsettling changes were sweeping TRG that went beyond the merger with Control Data to the larger world of corporate research. Still relatively small by comparison with companies like AT&T, Hughes, and Raytheon, TRG had been able to compete for research contracts on the strength of ideas that promised technological innovation. But the company, although it was making a few lasers, had no real production capacity. Increasingly, the larger companies recognized that they could underbid on research and still make money by winning the manufacturing contracts that could follow. Control Data, meanwhile, expected TRG to figure out how to use lasers to improve its computer-making capabilities, but didn't want to underwrite the research or wait for the results.

The research staff began to opt out of the confusion.

Steve Jacobs had been the first to go. After the successes of the cesium-helium laser and optical heterodyning, Jacobs felt the cutting edge was gone from TRG's research. He looked around and learned that Aden Meinel, who had headed the Kitt Peak National Observatory before moving to the University of Arizona to head the Astronomy Department there, was setting up an Optical Sciences Department. Jacobs applied for an interview and was offered a full professorship. He left TRG in July of 1965 for the desert heat of Tucson.

The University of Arizona was not the only institution that was getting into optical research. The laser had revived the study of optics to a remarkable degree. It was the hottest scientific field outside of space flight, producing a cornucopia of grants and a crush of student applicants to research-oriented physics programs. Brooklyn Polytechnic Institute had one such program. Brooklyn Poly's microwave institute, active in military research during World War II, had evolved into an Electro-Physics department that combined physics and electrical engineering. But the microwave radiation, propagation, and diffraction that had been its experimental mainstays were giving way to newer science. Administrators, smelling big research contracts in the field, wanted to develop optical—that is, laser—capabilities. They had submitted a contract to ARPA proposing optical studies. When Brooklyn Poly won the contract, it quickly had to find the staff it needed.

The institute's Long Island Center, its graduate campus where the high-level research was conducted, was at Farmingdale near the Republic Airport. TRG, which had moved its offices and laboratories to Huntington in 1963, was a stone's throw away. TRG's scientists, many of whom Gould had lured from Columbia and other research laboratories, comprised a nexus of doctorate level talent that fit the Polytechnic Institute's needs exactly.

Paul Rabinowitz had heard about Brooklyn Poly's plans early in 1965. By the summer, he had negotiated a job as a research associate that would allow him to take courses toward his doctorate while he helped the institute set up optical laboratories and a research capability at the Farmingdale campus.

Goldmuntz protested frantically when Rabinowitz and Jacobs turned in their resignations. He and TRG's other directors sensed that a few critical departures could build into a groundswell. And soon thereafter, the National Science Foundation inaugurated a program of grants to science institutions that accelerated the talent drain from TRG. The grants, of several million dollars each, were designed to help schools develop areas of new expertise in science, and to employ new faculty in those and traditional areas. Brooklyn Poly landed one of the grants, and Rabinowitz started lobbying to bring in scientists, including Gould, from his old employer.

■■

President Lyndon Johnson summoned Ali Javan and Theodore Maiman to the White House in April 1966. In a ceremony in the Cabinet Room, Johnson hailed their "brilliant and imaginative" achievements as developers of the laser beam, and awarded them the first Fannie and John Hertz Foundation Award, named after the rental car founder and his wife.

That summer, after returning his boat from its winter home in Antigua to Long Island, Gould slept aboard and served as a de facto night watchman in exchange for a free berth at its Huntington marina. Sailing occupied much of his time and thoughts as he waited for the Court of Customs and Patent Appeals to rule on his appeal.

August 4, 1966, was a Thursday. Gould was at work, looking foward to the weekend and sailing, when the phone rang on his desk. He picked it up and held his breath when he heard Bob Keegan's voice. He had prepared himself to hear bad news, but held out hope for good.

"Well, Gordon, I told you not to put any champagne on ice," Keegan said in the Oklahoma accent that years in New York had never managed to erase.

He said the CCPA had denied Gould's appeal on both conception and diligence, and awarded priority to Schawlow and Townes. The five-judge panel said Gould's notebook was "susceptible to numerous interpretations," and was "too ambiguous to justify the conclusion that the application possessed a definite and permanent idea of the complete and operative invention."

At the heart of the conception ruling remained the issue of the side walls. Gould essentially lost the case on conception by not having stated they were to be transparent, even though the medium inside was to be excited by optical pumping. Without that, his description was not enabling in that it would not have allowed someone who knew the field to build a laser. The panel went on to agree with the earlier ruling, and the Bell Labs lawyers, that Gould had not proven diligence, and that his own testimony was insufficiently corroborated. Ruth, the court said, understood what he was doing "only vaguely."

"Now what?" Gould asked Keegan. "Can we appeal?"

"There's no right of appeal. All we could do is ask the Supreme Court for a writ of certiorari. But the chance of them taking it up is somewhere between slim and none in a patent case like this." Keegan drew a breath. "We've got the others pending, and there's a lot left that the Schawlow and Townes patent doesn't cover. Don't forget, this was only your small application that was in interference. I don't think it's worth it to pursue this one any further. The company's not going to want to spend the money on the legal fees, for one thing." He paused. When Gould didn't respond, Keegan read disappointment into his silence and so offered him a straw of hope. "But why don't you think about it, and I'll talk to Larry, and we'll decide in a week or two." He knew, however, that the decision already had been tacitly made.

The Washington Post announced the result the next day with the headline "'Attic Inventor' Loses Fight For Patent Rights to Laser." The paper played up the quixotic angle. The story called Gould a "graduate school drop-out with a notebook of diagrams notarized in a Bronx candy store," and Townes "a Nobel

Prize-winning MIT provost." It referred to Gould's seven-year battle, and placed the stakes at "millions—perhaps billions—of dollars from applications of the laser."

That weekend was quiet aboard the *Wonny la Rue*. Gould drank more than usual, and smoked his usual three packs of cigarettes a day. By Monday, Goldmuntz confirmed what Keegan knew already. TRG would continue to pursue the other interferences, but Gould would have to give this one up as lost.

■

Gould found a level of solace in sailing. It was the kind of absorbing activity that allowed him to forget that his future lay in the hands of patent examiners and judges, and that his claims were opposed by ranks of well-paid corporate lawyers. There was always something to do aboard a sailboat. Surviving at sea required one's full attention and left no room to dwell on the quirks of the patent system.

He also had Brooklyn Poly's offer to consider, an offer that dovetailed with a plea from a group of TRG's other laser scientists that he take the lead in finding them a better situation. He could take the entire group to Brooklyn Poly if he chose. Gould remained loyal to Larry Goldmuntz, but Goldmuntz's days under the Control Data regime were clearly numbered. Gould decided to consider his options during an ocean passage to *Wonny la Rue*'s winter home in the Caribbean.

Gould and a small crew left Huntington during the last week of October. He paused at City Island, a part of the Bronx that is a boating center, for final provisioning. On October 29, a Saturday, a party of six left City Island and motored down the East River. Aboard were Gould, Marilyn, Jack Kotik, Bob Jeffries, a Connecticut sailor and technology entrepreneur Gould had met cruising on Long Island Sound, and Hans Hoff and Wolfgang Koenig, the boat's captain and cook, respectively. Marilyn went ashore at the Twenty-third Street boat basin. She waved as *Wonny la Rue* and her crew moved downstream toward New York Bay, the Verrazano Bridge, and the open ocean.

The predicted good weather deteriorated after nightfall the next day. Soon a fifty-knot wind lashed the boat and forced all hands on deck to reef the sails. The wind refused to diminish with daybreak. It was ripping the tops off the waves. The seas built until *Wonny la Rue* was plunging over thirty-foot swells coming from the quarter—at an angle to the stern—that made the full-keeled boat almost impossible to steer. On the second night, Hoff ordered sail reduced to a storm jib and a fully reefed mizzen, and that only to maintain control of the boat in the wild seas. The storm continued into a third night, and the seas continued to build. Hoff and his mate, with Gould at the wheel, took down the remaining sails as darkness fell, and the boat lay to in the heaving ocean.

Gould was in the cockpit with Hoff after midnight. The wheel was lashed to maintain steerage, and the two men were clipped to the boat with safety harnesses. Suddenly Gould heard a roar that the wind did not account for. He looked at Hoff and Hoff looked back and they both tightened their grip on whatever part of the boat they could grasp as a giant wave with a curling crest bore down upon them and descended upon *Wonny la Rue*. The wall of water pushed the boat onto her side and forced her masts into the water while Gould and Hoff dangled from their safety lines. It ripped the dinghy from the cabin top where it was lashed. Gould's first thought was that the water would crash through the cabin windows. Then he wondered if the boat would right itself, or whether he would die while so much in his life was unresolved. He had the odd thought that he would die blind, for the water had torn his glasses from his face; they hung from a cord around his neck. Gould could see nothing but black water and a smudge of white deck.

Wonny la Rue lay abeam the waves and each new wave pushed her back down as she tried to right herself. Finally, a lull between waves allowed the weight of the keel to take over and the yacht dragged itself upright, lifting the masts and rigging from the clinging waves. Hoff scrambled to the wheel and turned the boat into the waves, while Gould patted his chest to find his glasses and restore his sight.

The wind and waves diminished from that point, and soon Gould and his shipmates were sailing on in perfect weather. Well out to sea, far from comprehensive radio newscasts, the crew missed the announcement that the 1966 Nobel Prize for physics had been awarded to Alfred Kastler, whose work in optical pumping of atoms had transformed Gould's thallium experiment and started him thinking about the laser.

At another level of science, two weeks into the voyage, they encountered helicopters and warships, and learned they were skirting the area where astronauts Jim Lovell and Buzz Aldrin were expected to splash down on November 15 at the end of the final Gemini mission.

When *Wonny la Rue* made safe landfall at Antigua, Gould had reached a decision. Five months later, he said his goodbyes to TRG and joined Brooklyn Poly's Long Island Center as a full professor, taking with him the heart of TRG's laser research team.

30

Gould's return to academia followed by a day, coincidentally, a rare victory in his patent wars. The Board of Patent Interferences, on May 14, 1967, ruled for Gould in his interference against Fox and Kibler of Bell Labs over the use of Brewster's angle windows to reduce reflection losses in a laser. This, for a change, was a clear-cut, open-and-shut case in which Gould's priority was obvious. The lawyers for Bell Labs signaled that they intended no appeal.

Gould was one of six scientists who joined the Brooklyn Poly faculty that May in a simultaneous exodus from TRG. He still had no Ph.D., of course. This was unusual for a full professor, more so given that some of those who came with him, who had their doctorates, had lesser appointments. Gould remained the innovative thinker to whom people turned for new ideas. But his worth to Brooklyn Poly lay less in his inventive brilliance than in his ability to capture research grants.

Rabinowitz was still busy developing the Farmingdale campus's optical research capabilities. He had sold Gould to his colleagues as someone who could come up with projects designed to win grants that would put the new laboratory facilities to work. Peter Franken, who had testified for Gould in his effort to win a security clearance and as an expert witness in the Schawlow and Townes interference, had moved from the University of Michigan to Washington as deputy director of the Advanced Research Projects Agency, and had since become its acting director. Gould could count on Franken to look at any project he proposed.

Gould's first pitch to ARPA was a scheme to use lasers in a perimeter defense system. America continued to pour troops into the war in Southeast Asia, and the ease with which the stealthy jungle fighters of the Viet Cong could approach and attack vulnerable American positions was both frustrating and frightening.

Gould proposed, for the seond time, the use of a retroreflector in a laser system. This time, however, the retroreflector was the human eye. The "redeye" of people looking at the camera in old flash photographs was the reflection of the flashbulb off the retina. Gould believed these reflection capabilities, which were better than a cat's eye's, could work with a laser beam. He proposed to aim an amplified but invisible laser beam along a perimeter zone. When an enemy soldier penetrated the zone and looked into the path of the unreflected beam, his eyes would glow, signaling a security breach.

Franken squelched the proposal, but wouldn't tell Gould why. He couldn't say that a similar project was under way, this one proposing the use of neodymium YAG lasers not for perimeter security, but to blind intruders. The project was classified, and Gould still had no clearance.

Gould was not just a grant magnet at Brooklyn Poly. He taught courses, and advised members of the small cohort of graduate students in electrophysics. At least some of them found that Gould possessed the same ability he had seen in I. I. Rabi at Columbia—a way of opening their eyes to the processes at work. Bob Chimenti had completed his course work and was in the process of choosing a thesis topic about the time Gould arrived. Chimenti was interested in the still new field of lasers.

"He had an ability, more than anyone I had ever worked with, of having this physical insight," said Chimenti, who came to Brooklyn Poly with a master's in physics from Penn State. "He had a way of being able to explain things with an intuitive physical understanding of what went on in the processes of lasers. By the time you get to the Ph.D. level, you're used to having things explained the traditional way. But aspects of cavities and resonators, for example, whereas others approached them from a mathematical point of view, he could impart a different kind of understanding. He explained things in a really useful way, a way that clarified for people the physical properties and really made you understand it. And made you see new possibilities. It went well beyond the normal skills of a scientist."

Chimenti eventually chose the copper vapor laser as his thesis experiment, and Gould chaired the committee that oversaw his work.

■■

Charles Townes's fortunes also changed in 1967, though in a different way from Gould's. Townes fully expected to be named president of the Massachusetts Institute of Technology. But he failed to reckon with the doughty presence of a figure who had loomed over MIT for almost sixty years.

Vannevar Bush was an MIT Ph.D. who headed its school of engineering before World War II. As the war approached, he left Boston for Washington as president of the Carnegie Institution. He helped establish the National Defense Research Council, and during the war came to fame as director of the Office of Scientific Research and Development, the agency that oversaw the production of the atomic bomb. Bush, a politically conservative New Englander described by Richard Rhodes in *The Making of the Atomic Bomb* as looking "like a beardless Uncle Sam," returned to the Boston area in 1955 and was elected chairman of the MIT Corporation in 1957 and honorary chairman in 1959.

Bush's role as an advisor to five MIT presidents "was truly unique and without parallel," the corporation would say on Bush's death in 1974.

The problem for Townes was that Bush opposed an aspect of the nation's space program that Townes championed. Bush thought it was costly and inefficient to send astronauts into space when science had other pressing needs. He was against the lunar landing that NASA was planning to attempt before the end of the decade. Townes, by contrast, remained chair of NASA's Science and Technology Advisory Committee for Manned Space Flight.

Bush and Townes both were legends in their own time, but Bush was legendary at MIT, where it counted. No institute director or trustee would go against him. Bush's opposition was fatal to Townes's ambitions, and in 1967 he left MIT to accept an appointment as University Professor of Physics at the University of California at Berkeley.

■■

Research labs around the country continued to produce evidence that Gould's original predictions for the laser had been farseeing indeed.

In January 1968, two scientists at Varian Associates in Palo Alto received a patent on a method that allowed articles to be etched or coated by using a laser's heat to cause chemical reactions.

July 20 brought news of a patent granted for a laser made of carbon dioxide and noble gases. Five days later, IBM announced that two of its scientists had developed lasers that could be built for $25 to $50, using commercial dyes to generate their beams.

Larry Goldmuntz exited TRG that summer, leaving an unrecognizable company behind. Even Gould's old nemesis, Dick Daly, had left the year before to start a company that would manufacture ruby and other solid-state lasers. The laser business was outstripping the research business that had produced it. On the West Coast, Ted Maiman was another laser physicist who had turned to business, starting a laser company called Korad. Scientists like Maiman and Daly realized that an invention like the laser, with the promise of wide-ranging applications, came along rarely, and they decided that instead of looking for the next new thing, they would perfect and apply this one, for there was money to be made.

Control Data, with 60 percent of future royalties at stake, had continued to pay Darby and Darby for Keegan to pursue Gould's two remaining interferences, those against Javan and Hellwarth. Joe Genovese headed Control Data's patent department. While lasers remained a small and relatively unimportant part of Control Data's business, Genovese assured Keegan and Gould that the company was committed to seeing the interferences through.

■

More than ten years into his patent wars, Gould was no longer the naive and unsophisticated patent neophyte. He was beginning to learn the keys to the system, and finding that it worked in ways unimaginable to him at the beginning.

In 1968, the expanding Patent and Trademark Office had left its old home in the Department of Commerce Building and moved across the Potomac to sprawling new offices in Arlington, Virginia's, Crystal City. By that June, Gould's original two patent applications—the larger and more comprehensive one and its smaller counterpart designed to avoid classification—had been fileted through divisions and continuations-in-part into nine applications. The smaller one had given birth to two offshoots, both duplicated in the larger, and then had been abandoned a year earlier. The larger application, its "secret" classification lifted by the Air Force in 1967, remained alive and its April 1959 filing date the basis for the applications that had grown out of it. These covered not only the basic optically pumped and gas discharge lasers and their amplifiers, but the Q-switch and other components, the optical heterodyning scheme he had invented with Jacobs and Rabinowitz, and other laser permutations.

Not all of them were useful, except to provide tactical advantage against other inventors who had come along later with similar claims. Examiner

Ronald L. Wibert had had Gould's applications from the beginning. He had proven to be a sympathetic and thorough examiner, and he had tried to consider Gould's priority amid the welter of laser applications that had begun to flood the Patent Office. His problem, though, was that patent applications were kept secret in the United States until a patent issued. Since no patents had emerged from Gould's pioneering applications, Wibert had nothing to cite against newer claims.

Gould had filed foreign applications as well, however. America's defense alliances with the United Kingdom, Australia, and Canada had permitted the secret application to be filed in those countries, while the smaller application was filed in other European countries and Japan. Belgium had issued the first patent in 1962.

Sometime after that, Keegan had dropped by Wibert's office in the Department of Commerce building to discuss some issue pertaining to Gould's claims when Wibert directed his attention to a tall stack of papers standing on a cabinet.

"You might be interested to know what those papers are," the examiner said.

"I guess I would," Keegan said.

"They're copies of your client's Belgian patent," Wibert responded. He was using the disclosure printed in the Belgian patent, based on the same claims Gould had lost in the Schawlow and Townes interference, to reject more recent claims that duplicated Gould's. "I had five hundred printed," he said. "I've used about half of them so far."

That experience had prompted Keegan and Gould to do something to get his application printed and published in the United States. One of Gould's divisional applications, filed in 1967 after the larger application was declassified, made only some very remote claims for generating X rays and other higher-than-light frequencies. But it contained all the information in the original. Gould didn't care that he was patenting a device no one was likely ever to use. All he wanted was to get a patent so his original disclosure would be printed and published by the Patent Office. The patent, No. 3,388,314, was awarded on June 11, 1968.

Gould now had an additional bulwark in the Patent Office against claims that duplicated his. He also used the patent to restate his claim of priority over Javan, this time based on his earlier original application date rather than his notebooks and the material he had prepared for the TRG proposal. Nevertheless, the interference case went forward.

■■

Two Boards of Patent Interferences, composed of the same three examiners, had already begun reviewing evidence in Gould's two cases. Briefs and depositions had been filed earlier in the year by Keegan and his opposing numbers at Bell Labs and Hughes.

Keegan had argued in both cases that Gould should be awarded priority over the other inventors, since his was the earlier application date in both cases. The issue now was another nuance of patent law—undue experimentation. Javan and the Bell Labs lawyers argued that while Gould's patent application may have disclosed a laser excited by collisions of the second kind, it had not provided enough information for one "skilled in the art" to build one without an impermissible amount of trial and error in the laboratory.

They attacked Gould's application by turning its thoroughness against him, arguing that Gould had suggested gas discharge lasers operating by collisions of the second kind only as possible alternatives to other types of lasers, and pointing out that none of the mixtures he mentioned had been made to work at TRG. Gould had mentioned helium-neon, but did not suggest it was preferred, and did not mention excitation currents or powers, the Bell Labs argument continued, and only after two years spent discovering the right conditions had Javan and Bennett been able to make a working laser. Further, Gould had not given the basis for the physical dimensions of the laser tube he proposed, nor did he teach how to align the mirrors, and his calculation of losses in a tube 100 centimeters long would discourage one trying to build the device.

Bell Labs, again led by Torsiglieri, pounded especially hard on what it claimed were Gould's shortcomings in teaching alignment of the mirrors.

All of these arguments showed how devastating Gould's banishment from the actual experiments had been. While helium-neon and sodium-mercury mixtures were never tried at TRG, the promising krypton-mercury scheme was abandoned by Daly's experimental team even after the combination showed amplification that, at 1 percent, had proved sufficient at Bell Labs.

Gould's brief testified to the effects of his absence from the lab. It argued that his proposed gas laser tube—100 centimeters long and 1 centimeter in diameter—described operative dimensions, and that any worker in the art would have known the correct power values. Gould had not taught alignment of the mirrors because alignment using mercury vapor in a Fabry-Perot interferometer had been published long before. Further, he had included the loss-reducing Brewster's angle windows in his laser, which Javan had not, and that made it possible to place the mirrors outside the laser tube where they could be easily adjusted. Finally, Keegan argued that the only reason TRG had been unable to make a gas collisions laser work was because Gould "did not direct the project nor participate in the experimental work."

Here, Gould got a boost from Daly, of all people. Asked during a deposition why his group had not made the krypton-mercury collisions laser operate before abandoning it to concentrate on the cesium-helium optically pumped laser, Daly confessed, "Because we didn't build it the way Gould told us to."

■■

Hughes's argument echoed, in part, that of Bell Labs. They were close enough that Keegan believed they had shared notes and information as they sought advantage against Gould. But the Hughes lawyers put a bizarre twist on their case, and it could not have made the team from Bell Labs happy.

Hughes, like Bell Labs, conceded Gould's earlier filing date. They also conceded that Gould had disclosed the Q-switch for which Hellwarth had filed a patent application two years later. Hughes claimed, as had Bell Labs, that Gould had not disclosed an operative laser. But Hughes went on to argue that no operative laser had been invented when Gould filed his application. Therefore, he could not have invented the Q-switch since the Q-switch could not exist without the laser.

You had to think a minute before the implications of this were clear. Then its audacity was stunning, less for what it argued about Gould than what it implied about the original Bell Labs patent. Hughes's argument said the Bell Labs patent didn't work, that you couldn't pump potassium with potassium, that the Bell Labs "optical maser" was a dud. Hughes was saying that no adequate description of a working laser had existed before Maiman had built his.

If that was the case, Charles Townes and Arthur Schawlow had not invented the laser after all.

31

The interferences were still undecided late in September. Gould was sitting in the library at the Farmingdale campus one day, leafing through a copy of *The New York Times* when his eye fell on the first page of the business section. Suddenly, he went bug-eyed with disbelief. The headline that jumped out at him read, " 'Optical Pumping' Laser Process Is Patented."

Gould read, astonished, that the Patent Office had the day before—September 24, 1968—awarded a patent to the Westinghouse Electric Corporation "covering the invention in the late nineteen-fifties of a mechanism basic to the functioning of many types of lasers.

"The mechanism, invented by Dr. Irwin Wieder, uses optical means to stimulate the emission of light energy in liquids and solids such as rubies, glass and plastics. About half the lasers manufactured today are of these types.

" 'Optical pumping,' as the process is called, uses ordinary light waves to boost electrons or other particles in materials to higher-than-normal energy levels. This is a fundamental operation in producing the high-energy beams of light characteristic of the laser."

The story went on to say that the patent application had been filed in 1959, "the year before the word 'laser' was coined." When in 1959? The question sizzled in Gould's brain as he kept reading. Dr. W. E. Shupp, the Westinghouse vice president for research, had said the patent "is perhaps the most important issued to Westinghouse since that granted in 1937 for the ignitron," a device used to switch large amounts of electrical current. There was a photograph of Wieder, bespectacled and almost bald, with another researcher, both wearing jackets and ties in the manner of scientists dressing for the camera, peering at a helical lamp around the neck of a device shaped like an upside-down wine bottle.

By now Gould was jumping out of his skin. The Patent Office obviously had overlooked the Wieder application when it scanned laser filings for conflicts with Gould's application. Otherwise there would have been an interference. The conflicts here were glaring, if Gould could believe what he was reading. He dialed Bob Keegan's number with shaking fingers.

Keegan already had seen the story when Gould reached him, and was asking the same question: What was the filing date? A few days later he had learned that Westinghouse had filed the Wieder application on May 28, 1959, more than a month after Gould. It claimed an "optically pumped maser and solid state light source for use therein." The Wieder application had not applied strictly to lasers; it really was a long wave radio application, but he had thrown claims into the filing that allowed it to read on shorter wavelengths down into the infrared.

Keegan made plans to formally request an interference. But it was clear that Gould would soon enough be fighting on another front, this one potentially the most important.

■■

The Board of Patent Interferences released its rulings in the Javan and Hellwarth cases simultaneously on December 17. Keegan had left the city with his family for a Christmas vacation in upstate New York. He was grateful for that, because when he heard the news he was glad he didn't have to deliver it to Gould.

The three-member panel agreed with both Bell Labs and Hughes that Gould had not disclosed a working laser.

George W. Boys, La Verne Williams, and Stephen W. Capelli, in their opinion in Javan, toyed with the question of what a person "skilled in the art" might have known about making a laser in 1959, when Gould filed his patent application. The qualifications are "nebulous" with respect to a "new art which combines molecular physics and optics," the examiners wrote.

They ended up relying on the simplest of evidence. They said that because Maiman, at Hughes, had taken a year and a half to make his laser work from the time that Schawlow and Townes had published their article in *Physical Review,* and because Javan succeeded with the helium-neon laser only after two years of work, while TRG did not succeed in making a gas discharge laser operate, Gould must not have revealed a working laser. Neither, for that matter, had anyone else.

The testimony of all the witnesses "is not of much aid since the bulk of their knowledge in this field was acquired subsequent to April 1959," the examiners wrote. "However, their testimony coupled with the amount of experimental research in the three instances cited above, two which succeeded and one which failed, convinces us that the Gould application does not have an adequate disclosure to support the count as of 1959."

In other words, because TRG was unable to build a collisions laser, Gould could not have invented one.

The board went on to order that both Bennett, whose notebooks had been identified during the interference, and Donald Herriott be added as co-inventors with Javan, and awarded them priority.

In the Hellwarth case, the board reiterated that Gould had not provided sufficient disclosure. "The testimony of all the witnesses is fairly consistent in recognizing that the Gould application does disclose various working mediums, dimensions and theories," the examiners wrote, "but there is no information which is complete enough for one skilled in the art to build an operative laser particularly at a time when one 'skilled in the art' was virtually non-existent." Therefore, it ruled, Gould could not have invented the laser modification that allowed it to be "Q-switched" to produce a giant pulse. He could not invent an improvement of something that hadn't been invented yet.

Regardless of what it did to Gould, the board's agreement with Hughes's lawyers was a blow to Townes and Schawlow. Practically, it changed nothing. Bell Labs wasn't licensing their patent, and Gould had already lost that interference on other grounds. But the ruling reopened a door in the history of science that allowed the question to be asked anew: Who invented the laser after all?

Keegan had seen some bizarre rulings in his years as a patent lawyer, but this one was right up there, and he made plans to appeal. To avoid having to fight Bell Labs and Hughes at once, he decided to take the Hellwarth decision to the Court of Customs and Patent Appeals and Javan to the U.S. District Court in Manhattan. That, he hoped, would stop what he called "the double-teaming" on the other side. He also looked forward to a jury trial with live witnesses; he was tired of arguing points from dry and airless depositions.

▮▮

Gould was dispirited in the wake of the rulings. But this only accelerated what seemed to be a mid-life and mid-career crisis. Friends who worked or otherwise spent time with Gould as the 1960s waned noticed that in addition to his appalling cigarette consumption—still an unmoderated three packs a day—he drank more than was good for him, and often descended into blue funks about his teaching duties and the slow pace at which he was obtaining research grants. The depression of his college days returned, asserting itself as creative ennui. His only escape was weekends on the boat with Marilyn, weekends they spent cruising around Long Island and up and down the New England coast. It was only on the water that he came alive again.

Not the least of the reasons for Gould's dolor was the state of American politics. His old liberal instincts were offended by the times—by the continued escalation of the war in Vietnam, the seemingly unending toll of dead and wounded drawn largely from the lower rungs of American society, the succession of inept and corrupt governments those men were dying for, the atrocities. The assassination in Memphis of Dr. Martin Luther King, Jr., on April 4, 1968, struck at the heart of a cause he favored; Gould had been one of the early financial contributors to the Student Non-Violent Coordinating Committee and believed strongly in the civil rights movement. Robert Kennedy's June 5 assassination in Los Angeles, then the Chicago police attacks on protesters at the Democratic National Convention, and the Republicans' nomination of Richard Nixon in Miami, all helped thrust Gould into a swamp of mourning for his cherished principles.

His father's death on March 12, 1969, deepened his feeling of mortality and time passing. Kenneth Gould had retired as editor-in-chief of *Scholastic* in 1960, having added dozens of educational publications to the publisher's roster and expanded into textbooks, recordings, and other educational aids. He had remained active in retirement, editing a series of paperbound texts on world affairs for *Scholastic's* parent company, co-chairing the Workshop for Cultural Democracy, and advising the Scarsdale Adult School. His adult poetry class, taught nights at Scarsdale High School, drew packed audiences for his speaking voice, and the poems he recited from memory. But in the last two or three years he had grown frail. When he died peacefully in his sleep of a heart attack it was not a great surprise.

Kenneth, a social scientist, never understood the complexities of his first son's insight or the significance of his achievement. Gould had been over forty when his father presented him at Christmas with a book entitled *How Things Work,* and he would never forget his shock and disappointment not only at his father's primitive notion of science and scientific laws, but of Kenneth's utter failure to grasp what Gordon had done in inventing the laser. It simply wasn't real to him, Gould realized. His father was no different from the rest of the world. None of the signposts that people recognized were there—no patents, no awards, no mention in the history books. Somehow that reinforced his determination to keep fighting.

Helen Gould, three years older than her husband, remained in good health and stayed on alone at the house at 1 Berkeley Road in Scarsdale.

Gould continued to put in his time at Brooklyn Poly, but his grant proposals, which he had never been prolific in producing, were no longer certain of sympathetic review at ARPA. Peter Franken had resigned from the agency in January 1968. Gould's trouble landing grants was becoming a source of annoyance to his employers.

He kept *Wonny la Rue* in charter just enough to claim the boat's expenses as a tax deduction. He and Marilyn took their time aboard on legs of the cruises back and forth between Long Island and the Caribbean, and now and then spent a week sailing with friends out of the boat's winter port in Antigua.

Sailing companions recalled evenings at anchor. A day of sailing, or perhaps snorkeling over reefs inhabited by brightly colored fish, would typically end at anchor in a quiet cove. Gould, Marilyn, and their guests would have drinks, and wine with dinner in the cockpit under the stars. Conversation would meander the byways of science, politics, and music. One by one, the sailors would retreat to their bunks under the effects of the day's activities, the food, and alcohol. Gould was always the last one in the cockpit, still drinking, looking at the stars.

Jack Kotik sailed with Gould more than once. "He had trouble sleeping," Kotik said. "He mentioned once that his mind had an uncontrollable tendency to stay active, even after a long day, and he was probably trying to put himself to sleep. But in the end, it was always Gordon and a bottle."

32

Gould's landlords at the East Eighty-second Street brownstone came to him early in 1969, asking if he wanted to buy the house for $50,000.

The building was run-down, and Gould declined the opportunity. The owners sold to a woman who kept the house for three more years and then sold it herself, tripling her money. The new owners didn't want a tenant, and Gould left the cozy garden apartment for a high-rise building on East Thirty-sixth Street, where the Queens-Midtown Tunnel pours its inbound traffic into Manhattan's streets.

The move introduced a new living arrangement. He and Marilyn Appel, now in her third year of law school at New York University, decided that they would live together, and she left her apartment on Sixteenth Street at the same time.

Marilyn added necessary structure to Gould's life. She was a good sounding board and partner. This time, there was no talk from Gould of marriage.

Ruth Gould by this time had left New York and taken back her maiden name. She had resigned from the Columbia faculty in January 1969 and accepted an appointment at York University in Toronto as a professor of physics.

Gould was now approaching fifty and looking down the road. He wondered what he was going to do with the rest of his life, and how he was going to live the way he wanted to live. He had money on his mind, and he concluded that chasing research grants for Brooklyn Polytechnic was not going to do it for him. The pursuit of his laser patents still was going nowhere, and it seemed to

Gould that he had failed in his early dream of inventing his way to wealth in the manner of his hero, Edison.

So when Melvin Cook came along, sprinkling entrepreneurial pixie dust, he found Gould susceptible.

Cook was starting a laser company in Paramus, New Jersey, a few minutes from the George Washington Bridge. He sought out Gould at Brooklyn Poly and offered him a job as a consultant. What he really wanted was to use Gould's name to promote an initial public offering, and he paid Gould $15,000 from the proceeds.

Holobeam, as Cook called his company, was pursuing ventures in several laser-related areas, but the sizzle was the notion that people would pay good money to have their conventional photographs turned into three-dimensional holograms. Once Cook had investors on board, he asked Gould to invent what he had sold them on.

Neither Gould nor anybody else could make holograms from a conventional photograph in two dimensions. A hologram contains more information than a photograph, and it can't conjure up information that's not there. Gould tried other inventions, but none of them appealed to his new patron.

Cook continued to use Gould's name, now to lure scientists and other consultants, among them Rabinowitz. Rabinowitz worked on a laser system for the military, which used near-infrared lasers in place of live ammunition during war games. His job was to determine how laser detectors—the devices that would record a "hit" on a target—would respond to actual use in the field.

Gould, however, plunged in more deeply than the others. He placed enormous faith in Holobeam's prospects, to the extent that it seemed to Rabinowitz as if he was neglecting his work at Brooklyn Poly. Dismayed, Rabinowitz tried a pep talk, urging Gould to undertake the kind of first-rate science they hadn't been able to do at TRG.

But his exhortations fell on tired ears. Gould told Rabinowitz that he felt he was aging, past his prime scientifically, over the hill. He had the feeling that he didn't have that many years left. He was tying in with Cook and Holobeam because his patent situation was looking bad and he desperately needed to find the key to his financial future.

But neither was Holobeam the horse to ride, as Gould learned when he realized Cook had no concept of scientific reality. Cook thought Gould was such a great inventor he could take a wish and make it reality. When Gould told him the limits of possibility, Cook lost interest and Gould's newest dream of inventing his way to riches dimmed.

∎∎

For six days in July 1969, America and much of what was then called the free world gathered around television sets and watched as three astronauts—Neil Armstrong, Buzz Aldrin, and Michael Collins—rocketed toward the moon to keep President Kennedy's promise to the nation eight years earlier. The lunar module settled onto the surface of the moon on July 20 and Neil Armstrong gave the country the phrase "The Eagle has landed."

Several hours later Armstrong became the first man to set foot on the moon. Aldrin followed, while Collins orbited the moon aboard the mother ship, Columbia. A stark, almost colorless picture of Armstrong's boot on the moon's dusty surface beamed back to Earth. His message buzzed with static and was difficult for the millions watching and listening to hear: "That's one small step for a man, one giant leap for mankind."

Gould marked the American moon landing more as a milestone in his own long journey than as a celebration. He, like Vannevar Bush at MIT, thought the $20 billion spent on the moon landing could have been spent in other areas. "A machine could do what they did better," he complained. "It was a wrong choice of how to spend money just to have a guy plant a flag on the moon. That kind of money can do a hell of a lot of science."

He called the moon landing an engineering feat, discounting the way such a feat gripped the imagination. Lasers, too, were now more engineering products than scientific feats.

∎∎

In 1970, Control Data decided that lasers and other aerospace ventures that it had picked up with TRG's contracts were too far afield from its core business. It put TRG's laser operations on the block.

The eventual buyer was Hadron, a Westbury, Long Island, company founded by one of TRG's Ph.D. scientists, S. Donald Sims. By early 1970, Hadron was producing a wide range of lasers and Sims wanted to round out his product line. TRG's low- and medium-energy ruby and neodymium glass lasers, a series of YAG lasers, lasers for biomedical applications, and a range of energy monitors and other laser accessories gave him what he needed.

The revised patent agreement that Goldmuntz had insisted Gould sign when TRG merged with Control Data stipulated that the laser business couldn't be sold without the patents. The buyer would have to pursue the patents, and pay Gould royalties. That was trouble Sims didn't want. He just wanted to sell lasers, not shoulder the burden of prosecuting Gould's patent application, which now was more than ten years old.

Control Data was committed to pushing the deal through. The executive in charge of the liquidation called Gould and asked if he wanted to buy back his patent rights.

Suddenly, Gould had the chance to reacquire what he had been so reluctant to part with in the first place. For all of the difficulties he had faced so far, he remained convinced that he would eventually win patents, and that they would produce royalties. The laser industry continued to grow almost exponentially. The patents Gould believed were rightfully his had become, in prospect, very large cash cows.

The problem was, he had no money to speak of. Much of his net worth was tied up in *Wonny la Rue.* He spent most of what he made. The cost of maintaining the yacht and paying its captain and mate for the eight months a year it was available for charter was a drain, as were the trips between New York and the Caribbean. The expenses offset his income tax, though his accountant had warned that the Internal Revenue Service would tolerate no more than five years of losses.

Marilyn, when she had worked at Columbia, had met a law student named Joel Mallin. Mallin was blustery, a fast, loud talker, and smart. They had become friends, and Mallin had become friends with Gould as well. The three often enjoyed drinks and dinner together. Mallin was involved in financial law, and Gould and Appel decided to seek his advice about what to do about the patents.

Mallin was in Geneva, Switzerland, when Marilyn tracked him down. He was representing Bernard Kornfeld of Investors Overseas Services, one of the first large mutual funds of its kind.

Gould explained that he thought the patent rights could be worth millions of dollars—if he ever got any patents.

"You may think they're worth millions, but Control Data doesn't," Mallin said. "If they thought they were worth anything, they wouldn't be getting out of the business."

"But down the road, there's no telling what they could be worth. Ten million dollars, or more," Gould said.

"Gordon, get this out of your mind," Mallin advised him. "You're an ethical guy, but you can't be worried that you're taking advantage. You can't treat it as being worth multimillions, because for one thing, you can't afford it. This is your opportunity to steal it back. If they're shutting the whole thing down, take advantage of it."

"So what should I offer?" Gould asked.

"Tell you what," Mallin said. "Offer a thousand dollars, and 10 percent of whatever you ever make out of it." He paused, as Gould was thinking that it sounded like a good deal. Then Mallin added, "Up to a hundred thousand dollars."

Control Data accepted Gould's offer right away. The company's executives and Gould signed the necessary papers, and Gould came away with his patent rights, and a cabinet full of paperwork and legal files.

On April 27, 1970, Hadron announced it had acquired all the operating assets of the TRG laser department and electro-optics product line of Control Data's Melville Space and Defense Systems Division, as TRG was known within the company. Terms of the acquisition were not disclosed, but the best deal of the day was Gould's.

■■

Now the burden of prosecuting his own patents lay fully on Gould's shoulders. The prospective legal fees were daunting. He asked Keegan to continue representing him, with a caveat. "I can't pay you until I get some patents," Gould told him.

"I'll run a tab," Keegan said.

Gould turned *Wonny la Rue* over to a yacht broker at the end of 1970. The boat's usefulness as a tax deduction had ended, and he could not afford to keep it otherwise. He used some of the proceeds to pay down his debt with Keegan.

The Hellwarth and Javan cases poked on for months, stretching into one year and then two.

While those cases continued, the patent interference system brought Gould a second victory. The Board of Patent Interferences, acting in the interference against Irwin Wieder and Westinghouse, threw out all but one of Wieder's claims on the optically pumped laser amplifier. The Court of Customs and Patent Appeals upheld the board's decision on June 27, 1972. This didn't give Gould a patent, but it gave him priority. That meant that the field was now open to establish himself as the inventor of the process that had triggered the first laser.

About this time, Keegan had begun laying plans to leave New York after nineteen years. He had decided—for reasons unfathomable to Gould—to move to Fayetteville, Arkansas, where he would become the only patent attorney in the state. Keegan was not leaving immediately, but when he did, he told Gould, he would continue to represent him until the remaining interferences were resolved. Gould's professor's salary didn't leave room for legal fees, and they both understood that Keegan would be working on the cuff again.

Meanwhile, Gould's tenure at the Brooklyn Polytechnic Institute was proving disappointing both to Gould and to his employers. He taught a regular course load, and he was a good teacher. He brought a manner to his classes that students seemed to enjoy as he stood at the blackboard with his ever-present cigarette in one hand and a piece of chalk in the other. His lectures were clear, and his breadth of knowledge unassailable.

His enthusiasm for a good invention was undiminished, too. Bob Chimenti and Rabinowitz together invented an image amplifier that employed Chimenti's copper laser. It used laser action to make an image more intense, without flooding it with light that could—as in the case of a tissue sample or an old and damaged piece of text—destroy the object that was being imaged.

When the two excitedly conveyed to Gould what they'd come up with, he was equally excited. "Hey, that's a neat thing. I wish I'd come up with it," he said, and Chimenti and Rabinowitz left feeling that they had won the approval of a real creator.

But even to Rabinowitz, who had received his Ph.D. in 1970 and stayed on to teach and do research, it was clear that Gould's interests lay elsewhere. Relative hard times had descended on the institute, which made matters worse. The prosperity conferred by the National Science Foundation with its $2 million Science Development Fund grant was replaced by a freeze on faculty hiring. Rabinowitz left the institute in 1972 to go to Riverside Research.

At midyear, Keegan and patent attorneys at Hughes Research—W. H. MacAllister and Paul M. Coble—filed their appeal briefs in Gould's interference with Hellwarth over the Q-switch.

The lawyers for Hughes leaned heavily on the testimony of Dr. Bela A. Lengyel, who was a laser historian as well as a physicist who had worked on lasers at Hughes. They cited as one of Lengyel's credentials the introduction to his 1966 book *Introduction to Laser Physics* by Charles Townes, in which the Nobel Prize winner lauded the Hungarian-born Lengyel for his success at completing a "terribly difficult and highly important" book.

At least, by then, Townes was content to let lasers be lasers; he dropped "optical masers" from his vocabulary in the introduction.

Lengyel's testimony, which the lawyers synthesized in their appeal brief, was a compendium of the experimentation that was done at laboratories like Hughes, Bell Labs, IBM, RCA Laboratories, and others, including TRG, before Maiman's laser pulsed for a fraction of a second. The lawyers quoted Arthur Schawlow's paper delivered at the first International Conference of Quantum Electronics in September 1959, when Schawlow said, "As yet, nobody knows for sure what form a practical source of coherent infrared or optical radiation will take."

The lawyers skewered Gould, in other words, for the failures of imagination that were widespread throughout the community of laser experimentation. They went on to recount "the dismal results of TRG's attempts to build a laser during the years following Gould's filing date." They admitted that Gould had invented the Q-switch, but they pressed hard their case that since the Q-switch needed a laser in which to operate, Gould could not be given the patent.

Keegan called Hughes's argument "extraordinary." "There is only one criticism that can be made of Gould's disclosure; it was filed before anyone had built a laser," he had written at an earlier point in the interference process.

Now he attacked "the audacity of the argument of Hellwarth that he should be entitled to a patent because Gould's description of his conception of an element of the combination [laser with Q-switch] in issue was insufficient, while Hellwarth did not conceive this element at all, but rather appropriated the concept from another."

Audacity it certainly was, and it was embodied in the bluster and gestures of the lawyers on the other side. The opponents convened at the federal courthouse in the District of Columbia to make their arguments in support of their briefs. The hearing ended, and as Gould was leaving the courtroom with Keegan he thought he saw one of the opposing attorneys raise a middle finger at him.

33

The axe didn't fall for several months. Then, on February 15, 1973, the Court of Customs and Patent Appeals rejected Gould's appeal. The decision reiterated the board's finding, pressed by Hughes, that Gould's patent application had not disclosed an operable laser. It "is not questioned," the court said, "that the disclosure of the Q-switching feature would be adequate if the application adequately disclosed an operable laser in which the feature could be incorporated."

The ruling said, in effect, that nothing before 1960, when Maiman operated the laser from which Hellwarth had devised his Q-switch, had enabled one "skilled in the art" to make an operable laser. That included Schawlow and Townes's patent application. The decision did not refer specifically to Schawlow and Townes and did not affect their patent, but the implication remained clear: if Gould had not described a working laser, then neither had they.

Gould might have felt a little better in defeat to think that Townes and Schawlow also were deemed by the decision to be inadequate. But he thought the court's logic was strange. It seemed to turn on its ear the practice of the Patent Office since the nineteenth century, that except for claims on the invention of perpetual motion machines, working models weren't required for patents to be granted.

Keegan asked for a rehearing. But Gould was, by now, familiar with the glacial pace of patent matters. The setback made him take a new look at his prospects, and when an old friend approached him with a business opportunity, Gould was ready to listen.

■■

He had already decided to leave Brooklyn Poly. The school was working on a plan to absorb New York University's engineering department and change its name to New York Polytechnic Institute. The combination would create a teacher overload, and so Brooklyn Poly was dangling attractive financial packages in front of tenured professors as incentives to leave the faculty.

By taking a "terminal leave of absence," Gould could walk away with a year's pay. That looked like a good deal, and he took it. When his resignation took effect on March 15, 1973, he already had started working with Culver.

Bill Culver, since meeting Gould in 1959 and being thrilled to find someone with such advanced laser ideas, had gone from Rand to the Institute for Defense Analysis. There, he had served under Charles Townes from the fall of 1961 until 1965. He had met with a series of laser investigation panels with names like the Ad Hoc Group on Optical Masers, the Special Group on Optical Masers, the Advisory Group on Electron Devices, and an unnamed study group convened to decide if the military should start a formal laser weapons program.

Although he worked under Townes, Culver had great respect for Gould's thinking and he routinely invited him to attend the laser group meetings despite his lack of a security clearance. "We'd invite Gordon down," he said. "And he'd come in and talk to us. But then he'd have to leave."

A facet of the meetings that Culver remembered was a Bell Labs presence, and a corresponding disdain for Gould's ideas. "Because there were a lot of Bell Labs people around who always wanted to put Gordon down, and since he never got a security clearance and couldn't stick around to defend his ideas, they never got to be popular," Culver said.

Culver went from the Institute for Defense Analysis to IBM, while Gould left TRG for Brooklyn Poly. Then, in 1971, Culver had the idea that he could make accurate short-range missiles using fiber-optic cable.

"We could hit Moscow with missiles launched in North Dakota, but we couldn't hit a tank on the other side of the hill," Culver said. "But with optical fiber, you could equip a missile with a television camera in the nose, and a spool of optical fiber, and pay it out, and it would be just like a kamikaze, except that you could be back somewhere and see what was going on. You don't miss your garage when you go home at night. There's no reason why, if you could see it, you should miss the target." Culver started going around talking about his rocket on a string.

Fiber optics were still relatively new. IBM wouldn't buy into Culver's idea.

Neither would the big defense contractors. But the Army Missile Command (MiCom) at the Redstone Arsenal in Huntsville, Alabama, told him it was a good idea, and in August 1972, Culver started his own company to pursue it. He called it Optelecom.

Optelecom seemed certain to win a contract from ARPA for Culver's guidance system. His plan was to sublet some of the work, and he thought of Gould.

In the optical fibers that existed then, the losses were fairly low over two kilometers. But the Missile Command had a five-kilometer requirement in its missile contract. Culver's idea was to use a laser on the ground to transmit a modulated signal to a missile equipped with a retroreflector in order to guide it to the target over the remaining three kilometers. Culver had chosen deuterated ammonia—ammonia in which one of the hydrogen atoms is replaced with deuterium—to modulate the signal. Its problem was corrosiveness. It ate up the stuff that contained it just as cesium gas did.

Gould agreed to join with Culver on the contract. He had already decided to leave Brooklyn Poly. The school had begun working on a plan to absorb New York University's engineering department and change its name to the New York Polytechnic Institute. The combination would create a teacher overload, and so Brooklyn Poly was dangling attractive financial packages in front of tenured professors as incentives to leave the faculty.

By taking a "terminal leave of absence," Gould could walk away with a year's pay. That looked like a good deal, and he took it. When his resignation took effect on March 15, 1973, he had already started working with Culver.

As soon as he was off the faculty, Gould and several other renegade professors turned around and rented laboratory facilities at the Farmingdale campus, where they started doing research in Culver's contract. In a matter of weeks, Gould replaced Culver's gas of deuterated ammonia with carbon-13 methyl fluoride, which was better and easier to work with.

Solving the modulator problem fulfilled Optelecom's first contract, and ARPA gave the company a contract to build an entire system. Culver then decided to incorporate Optelecom, and brought Gould in as vice president and co-owner. Culver got 51 percent of the stock, Gould 25 percent, with the remaining 24 percent divided among other employees.

Optelecom continued to rent out laboratory space at the Farmingdale campus, where Gould worked on the company's optical communications system contract. Culver tried to persuade Gould to transfer his operations to the Washington area, but Gould resisted. He and Marilyn wanted to stay in New York and maintain their lifelong love affair with the city. Every time he made one of his frequent shuttle trips to Washington, he tried to convince Culver their company needed a permanent branch office in New York.

■■

At the end of the third week of June 1973, Gould received Bob Keegan's letter telling him the Court of Customs and Patent Appeals had denied his request for a rehearing. The only appeal, again, was to petition the Supreme Court for certiorari, "a one to one thousand chance at best," Keegan wrote. "It appears, therefore, that we shall have to consider the Hellwarth interference closed as a practical matter."

That left Gould's box score in interferences at two for five. He had lost to Schawlow and Townes on the optically pumped laser oscillator, to Hellwarth on the Q-switch, and to Javan on the gas discharge laser, although the Javan case remained on appeal before the U.S. District Court for the Southern District of New York in Manhattan. He had prevailed against Wieder on the optically pumped amplifier, and against Fox and Kibler on the use of the Brewster's angle window. None of these decisions either granted or denied patents, they simply clarified the possibilities. Gould's original patent applications had now transmuted into divisional and continuation applications that kept his claims alive and separated each invention from the others, and it was time for him and Keegan to salvage what they could of his patent prospects.

Keegan thought some urgency was needed. His move to Arkansas was only six weeks away, and he was anxious to resolve some of the questions surrounding Gould's patents. "There is the matter of the controversy with Bell Telephone Laboratories and the possibility of salvaging some amplifier claims, and perhaps some other matters that should be discussed rather promptly," his letter to Gould continued.

"We shall have to advise the Court in the Bell Telephone Laboratories case of the final resolution of this matter, assuming that no petition is to be filed. If any proposition is to be made to Bell Telephone Laboratories, it should be done very promptly."

Keegan urged Gould that they get together "in no case later than the end of this month."

Keegan's urgency had only partly to do with his upcoming move. He believed he could settle the Javan case on terms that were favorable to Gould. He had dropped hints to Bell Labs' lawyers that a settlement was possible, and while they hadn't rushed to the bargaining table they weren't pushing the case, either. Nevertheless, Keegan understood that the case wasn't going to sit there forever, and with the Hellwarth interference lost, the time was ripe to push for a resolution.

The idea was to limit Javan's claims as much as possible, and preserve Gould's priority for what was left. Gould still had the earlier filing date, and if Keegan could manage a settlement that held Javan to narrow oscillator claims and the specific combination of helium and neon covered in his application, Gould would have much to claim patents on.

And Keegan, as he had said, also had some other matters to discuss.

■■

The inventor and the patent attorney met during the last week of June at Darby and Darby's Lexington Avenue offices. Keegan laid out his settlement ideas, and Gould agreed. Then Keegan urged Gould to accept a major refocusing of his patent claims. It was not the first time he'd brought up the question with Gould. The patent examiner, Ronald Wibert, had given him the idea four months earlier.

Keegan had talked to Wibert often throughout the extended interference processes. As the tide went against Gould again and again in the interferences, Wibert had told Keegan that he and Gould should make a choice. "There are two inventions here," he said. "Some call for an oscillator, and others call for an amplifier. You need to choose."

Wibert had first made the suggestion on March 8 in urging Keegan to file a divisional application on Gould's claims, now that they had been largely sorted out in the interferences. It had struck Keegan with the force of revelation.

The lawyer had written Gould the same afternoon. "Dear Gordon," he wrote. "I have a very interesting conversation to report to you.

"The Examiner called me this afternoon and asked me to make an election between two different inventions which he deemed to be involved in the application. The two different inventions were amplifiers and oscillators. It suddenly occurred to me that as to amplifiers, the CCPA [Court of Customs and Patent Appeals] decision is most likely not controlling. I don't think we would ever have thought of this possibility if the Examiner had not mentioned it.

"The Examiner said that he deemed the claims won from Wieder to be all oscillator claims, but I questioned his evaluation on this point. We went over the Wieder claims and he agreed that most of the Wieder claims are, in fact, amplifier claims. I told him that based on this reallocation of the claims, we would elect the amplifier claims."

Keegan went on to point out to Gould that the Court of Customs and Patent Appeals decision in Hellwarth did not affect amplifier claims, and added, "I think you will agree that this is certainly an interesting development and one which we will have to consider in our review of this whole situation."

The distinction Wibert had made was not an especially subtle one. A laser, like many machines, is an assemblage of components. One is the oscillator, the mirror-ended part of the device that makes the laser beam coherent. But the amplifier is what starts the process, the trigger that creates a population of excited atoms in the laser medium that can then be stimulated to emit their radiation and—as the new vocabulary that had grown out of Gould's original coinage had it—begin to lase. An oscillator without an amplifier was just a set of facing mirrors.

Gould clung to his oscillator claims like a particularly tenacious barnacle. He was proud of his amplifying schemes, but he also knew that being able to select and isolate a coherent beam was what made the laser the incredible tool it was. Although he had been rejected at every turn where oscillators were involved, he felt that the original concept of the opposing Fabry-Perot mirrors belonged to him and no one else.

Still, Gould at almost fifty-three had become a realist. That his original filing date and the unadulterated contents of his first applications had survived for more than fourteen years was extraordinary. Those applications had produced four minor patents, including the X-ray laser, but only the extended interferences had kept the remaining possibilities alive. These were the most important, because they were basic laser processes, in widespread use. If he refiled now to seek out the claims not denied him in the interferences, as Keegan proposed, he would still have the benefit of an application filed before most other laser applications.

Keegan finally convinced Gould to file a continuation application focusing on amplifier claims. He took the work with him when he headed south, and a year after he left New York filed the new application on August 16, 1974. The Patent Office gave it Application Serial No. 498,065, which was a continuation of Serial No. 644,035, abandoned, which was a division of Serial No. 804,540, Gould's original big application, and a continuation-in-part of Serial No. 804,539, the smaller of the original April 6, 1959, filings. It was also a continuation of the 904,540 application. The descent of third cousins at a family reunion was easier to follow.

The new application avoided claims that required a working oscillator in order to be patentable. Keegan instead focused on the optically pumped and gas discharge laser amplifiers, the uses of lasers that Gould had predicted, and the use of Brewster's angle windows to avoid reflection losses in a laser.

■■

Keegan, when he filed the latest application, had done nearly everything he could for Gould. He was still working virtually for nothing, with the informal understanding that if Gould's applications ever turned into patents and the patents produced royalties, he would be paid. But Keegan knew the way things worked. He could see the laser industry growing bigger every day, and his experience in patent law had taught him that if Gould did win basic patents, the industry would rebel at paying royalties on them. Three years earlier, in fact, laser makers had started paying royalties to Charles Townes and his patent agent, the Research Corporation, after Townes had sued Spectra Physics, the largest laser maker, to enforce his contention that his optically pumped maser patent was the basic patent that underlay the laser. The people who made and used lasers wouldn't welcome another inventor with claims on their pocketbooks.

Keegan had left New York to make his life easier. He was enjoying his new situation in Fayetteville. He didn't want to get involved in protracted litigation that would find him leaving his wife and six children at home while he spent weeks on end in courtrooms and hotel rooms in New York and Washington.

"Gordon," he said when they discussed the matter, "I'm a one-man band down here. I think you'll get your patents, and then I think you'll have to sue to enforce them. You need a larger firm to do that." He set a cutoff date and said, "I'll stay with you until then, and I'll still run a tab. But after that, you need to have somebody else."

That sent Gould looking for a new way to pursue his patents. He already had spent about $40,000 out of pocket, but that was well short of his total legal bill. He still owed Darby and Darby about $60,000 for Keegan's time, and he had no funds for a sustained legal battle. The only possibility he saw was bartering away a share of the prospective patent income. The question was, who would take a chance on income from patents that didn't, and might never, exist.

One potential partner had fallen by the wayside. Bio-Med, a medical laser company started by Gerard Grosof, Gould's former TRG colleague, had promised to provide some funding for the patent battle. But the young company withdrew, saying it didn't have the cash flow to support the substantial legal fees even while Gould was still arguing the interferences.

By the beginning of 1975 Gould, still working on Optelecom's contracts in Farmingdale, was tapped out. He had barely enough money to keep his latest application alive. His ears were sharp in his desperation, and when he heard that a wealthy entrepreneur had called one of New York Poly's professors to talk about a laser article he'd written, Gould got the man's name and tracked him down.

34

John Coleman had benefited from one of the steepest fifty-two-week stock climbs in pre-Internet history. He was one of the founders and principal owners of Radiation Research, a NASDAQ traded company whose stock rose from $5 a share to $500 in 1963 on the strength of its glow discharge coating patents that led to licensing agreements with major companies. (Glow discharge, or plasma, coatings are applied as a glowing gas and are used as protective coatings, today principally on computer chips.) The shares had fallen from their peak when Coleman sold out in 1967, but they still made him a wealthy man. He played polo and kept horses on his estate in Locust Valley, on Long Island, when he made the inquiry that led Gould to contact him.

Coleman's willingness to hear what Gould had to say grew out of his own background. He had entered an accelerated program designed to produce engineers for wartime service during World War II, but the war had ended when he graduated from the University of Virginia. After three months in the Navy, he went to work at the RCA Laboratory in New Jersey and started working on a Ph.D. in physics at Princeton at the same time. Princeton, like Columbia, was a hotbed of physics after the war. The last three lecturers Coleman heard were Niels Bohr, Albert Einstein, and Robert Oppenheimer, and the thrill of those lectures never left him.

After Princeton, he started Radiation Research in 1950, with two undergraduate friends. The company's spectacular fortune revealed to Coleman the value that patents could hold. "I was inclined to be sympathetic with a poor starving physicist," he said.

Coleman invited Gould to his Locust Valley estate. Gould entered a long driveway and drove through acres and acres of green lawn. He saw horses grazing behind white rail fences. Coleman obviously was a person of means. They talked, and Coleman found Gould's story compelling.

When Gould said that he'd been filing continuations and divisional applications with the Patent Office for years, but was almost out of funds, Coleman said, "I'll personally guarantee the money to pay your next filing fee just to keep your application alive."

Coleman then contacted Howard Turner. Turner was the chairman of the world's largest building construction firm. In New York alone, Turner Construction had built three of the Lincoln Center Buildings, participated in the construction of the United Nations, and built the Ford Foundation headquarters. The firm at the time was erecting some of Hong Kong's largest buildings. Its chairman, however, had a scientific background. He had a Ph.D. in chemical engineering from MIT, and had worked in scientific fields where he'd met Coleman. They were partners in a small research firm, Plasma Physics. Turner listened to what Coleman had to say, and agreed to meet Gould.

Coleman took Gould, accompanied by Marilyn, to Turner's office at the top of the Mobil Building on Forty-second Street across from Grand Central Terminal. Gould was stunned by the size of the light-flooded office. It dwarfed even the large oval table Turner used for a desk.

Turner and Coleman saw in Gould's situation a straightforward business opportunity. Turner was an alumnus of Swarthmore College, and through his service and contributions to his alma mater he had met another active Swarthmore alumnus, Eugene Lang. Lang, a few years younger than Turner, had developed and sold a series of successful small businesses before hitting upon a plan that would allow him to nurture several businesses at once. Refac, the firm Lang founded in 1952, was in the business of technology development. It helped inventors exploit technology by licensing their patents, primarily in foreign markets.

Gould at that point had patents in Canada—he had won an interference against Javan and Bell Labs there in 1972 over the gas discharge laser—as well as in Belgium and the United Kingdom. So he had patents to license; the biggest licensing fees, however, remained in the future, in the patents he had yet to receive in the United States, where the overall laser industry in 1975 was approaching $300 million.

Gould's impression was that Refac would help him obtain his U.S. patents.

█

Turner arranged the meeting, and he and Coleman accompanied Gould and Marilyn to the Refac offices, farther east on Forty-second Street. Lang was short and compact, with a bland face and shrewd eyes. The five of them sat on leather chairs and couches around a large square coffee table, and Gould repeated his story.

As Turner had predicted, Lang identified with Gould right away. He had grown up one of three children in East Harlem. After the family moved to heavily German Yorkville on Manhattan's East Side in the early 1930s, he remembered his father standing up to Nazi sympathizers who revealed themselves with cheers during movie newsreels of goosestepping German troops. That experience lay at the core of Lang's antiestablishmentarian feelings. He was impressed with Gould's persistence, his belief in himself and his invention, and especially the sense that the scientific and patent establishments were arrayed against him. He thought Gould had been shamefully dealt with from the very start.

"I was captured by an array of considerations," he said later. "First, I liked him. He was a simple, unpretentious guy. I think he had genius, because I didn't really understand the technology he was trying to explain to me. And, of course, he had kept his patents alive in the patent office against the opposition of many others."

Lang knew the rights to the laser were potentially extremely valuable. It was that value rather than any empathy he felt that led him to ally with Gould. And for Lang and Refac, an alliance was a risk-free proposition.

Lang said it had always been Refac's standard agreement to work at licensing patents—as well as proprietary technical and manufacturing know-how— for a 50 percent share of licensing revenues. "I made a conventional deal with him," Lang said, "where we acquired the exclusive world-wide right to license his patents. Our deal was that we would, entirely at our risk and expense, exploit his patents."

This fifty-fifty split between Gould and Refac, however, according to Lang never included Refac's commitment to help obtain the patents.

"He would remain responsible for prosecuting and maintaining his patents," Lang said. "We were not patent lawyers. We were not in the patent business. We were in the business pure and simple of creating new enterprise based on proprietary technology."

Refac, in other words, would not put one dime into prosecuting the U.S. patents that it hoped to license.

Gould, however, came away from the meeting with a different impression. He thought that for half of the potential rights to royalties on what Lang conceded had become a major worldwide industry, Refac would help him pursue his patents.

John Coleman received the same impression, that Refac would prosecute as well as license the Gould patents. Without patents, after all, there would be nothing to license.

The misunderstanding would lead to much future bitterness, but at the time Coleman remembered Gould as being ecstatic with the turn of events. Coleman thought his offer to absorb the filing fee for Gould's next continuation had given him a new dose of hope, although his generosity was relatively minor. Patent Office filing fees had always been kept low in the interest of encouraging independent inventors. In 1975 they were only $65, and while Gould's renewed patent application had been divided into several separate applications, Coleman's offer amounted to only about $250. In exchange, Gould had agreed to give Coleman a piece of his Canadian patent income.

But Gould was far from ecstatic about the finder's fee Turner and Coleman were demanding for recommending him to Lang. He and Marilyn had returned to Turner's office, where the construction magnate produced a single sheet of paper and placed it in front of Gould to sign. It was an agreement to turn over 10 percent of whatever income resulted from the meeting. Gould had nothing at that point, but having just signed away half of his future income, he resented a grab for another 10 percent just for an introduction. When he resisted, Turner tried to overwhelm Gould by insisting that he owed Coleman and Turner. Marilyn had never before run into such a force of personality, but they left without signing the paper, and Lang ended up compensating Coleman and Turner for bringing Gould to him.

■■

The immediate problem became keeping Gould's patent application alive. Keegan had agreed to continue helping Gould for as long as he could, but he had set a cutoff date, and the days were counting down.

Lang, while maintaining that Refac was not obligated to prosecute Gould's application, understood that without patents his deal with Gould was worth nothing. He knew a patent lawyer named Joe Littenberg, who had handled and won a case for Refac. He called Littenberg at his Westfield, New Jersey, offices and sketched out Gould's story. Soon Littenberg and one of his partners, Dick Samuel, were on their way to Manhattan in Littenberg's Cadillac, carrying a copy of Van Nostrand's *Encyclopedia of Science* to help them decipher the laser. By now it was the spring of 1976.

Samuel, thirty-six, bearded and intense, didn't have high hopes for the meeting. He had worked at Bell Labs as a patent lawyer before joining the law firm, and he knew that Bell Labs held the basic laser patent. He had called a friend who still worked there and asked if he knew anything about Gould.

"Oh, the sad claims of Gordon Gould," said Samuel's friend. "He's a nut, a kook. He's always claimed to have invented it, but we have the patents to prove otherwise."

They reached the Refac offices, and a receptionist ushered them back to Lang's corner office with its leather sofa and chairs, where a man with gray-flecked wavy hair sat smoking.

"This is Gordon Gould," Lang said.

Gould rose and stretched out a hand with nicotine-stained fingers. Samuel took him to be in his mid-fifties. His face had a sun-lined, weathered look, and the corners of his eyes were creased from squinting. He wore a suit and tie, but his soft-soled shoes betrayed him as preferring casual dress.

"I understand you invented the laser," Samuel said when everyone was seated.

"That's correct," Gould responded quietly.

"That must have been a long time ago."

"Yes, it was 1957."

"Did you file a patent application?"

"Yes."

Samuel, like most patent attorneys, was always running into people who claimed inventions from fire to the wheel. Talk to them, and to hide their fool-ishness they'd cloud the air with bullshit. Gould wasn't doing that. He exuded calm determination, and his eyes were alert and appraising. He was at least as interested as Samuel in deciding whether the other man was serious. Samuel began to be impressed, but was still wary.

"Is the patent application still pending?"

"Yes."

"Do you have the papers to show it?" Here was the key, Samuel thought. Most of the forgotten inventors had a story. Their dog ate the paperwork, it had burned up in a house fire, a flood had swept it away; there was always a story, but never a piece of paper to support their claims.

Gould reached down to a large, scarred, very old leather briefcase at his side, sorted through some papers, and handed them to Samuel. Samuel looked at them and saw they were official actions from the U.S. Patent and Trademark Office. He traced their descent through divisions and continuations and real-

ized with growing surprise that the file was current. The date on the most recent action, a communication from an examiner, was just a few weeks old. The
original filing date was in April 1959. Samuel retraced the history of the application in the papers before him; they had all the stamps, all the dates, all the
right indications. The application had a tortured history, but it was still alive.
He's not a nut so far, the lawyer thought.

Then he remembered his research. "But the Bell Labs application was filed
in 1958," he said. "If you applied in 1959, how do you expect to show priority?"

Gould leaned back to his briefcase. This time he produced a battered laboratory notebook, which he handed to Samuel, again without comment. Gould's
name was printed by hand on the cover and, beneath it, "Notebook #1."

Samuel opened the notebook. He could feel Gould watching him. The
handwritten heading at the top of the first page read, "Some rough calculations
on the feasibility of a LASER: Light Amplification by Stimulated Emission of
Radiation." Below that was a diagram around which Gould had written, "Conceive a tube terminated by optically flat partially reflecting parallel mirrors."
Some kind of stamp appeared in the left-hand margin, and Samuel turned the
book sideways to see what it was. It was a notary's seal, signed and dated November 13, 1957. The writing, interspersed with formulas and sprinkled with
footnotes, went on for nine pages, the last three devoted to a "brief statement
of properties and possible uses of the LASER." The notary's seal appeared on
all of them.

"Everything that's in that Bell Labs patent is in here," Gould said quietly.

"Are you saying this is the original description of the laser?"

"Well, I think it is," Gould said, almost wryly. "Nobody's ever proved it isn't."

"Why didn't you get the patent, then?"

"Well," Gould said with a grin and a smoke-roughened chuckle, "that gets
to be kind of a long story."

■■

By the summer, the partners of Lerner, David, Littenberg and Samuel had
agreed to invest up to $300,000 worth of time and expenses pursuing Gould's
patents, in exchange for a 25 percent share of future royalties. They created a
company to hold the interest, and called it Creative Patents, Inc.

Gould objected to giving up half of his remaining future royalties while
Refac, for its 50 percent, was only peddling his foreign patents and waiting for
the far more valuable American ones. He pushed Lang for a concession that

softened the basic Refac deal, reducing Refac's share to 40 percent. Gould agreed to cede the law firm 15 percent. When the deal was sealed, Gould was left with a 35 percent share, Refac with 40, and the law firm had a quarter.

The agreement was open-ended on the law firm's side. Lerner, David made no commitment to pursuing the patents to the bitter end, and had the option of withdrawing at any time. Samuel and Littenberg, as well as other members of the firm, would handle aspects of the case.

When the contingency agreement was signed, Samuel reviewed his understanding of the laser. He had an electrical engineering degree from Rensselaer Polytechnic Institute before getting his law degree at Boston College, and a master's in patent law from George Washington University. But neither they, nor the Van Nostrand textbook, had addressed the laser's full complexities. He needed a booster course. He sat down with Gould and said, "Okay. You have to explain to me, very basically, how a laser works."

When he understood it, he began a thorough case review, poring over the interference proceedings and talking at length with Bob Keegan, looking for an approach that would reverse Gould's history of failure.

Keegan told Samuel that Ron Wibert, the examiner, had urged him to prosecute amplifiers in the wake of the rejection of Gould's oscillator claims. Wibert had since retired, and the 1974 application now rested with an examiner named Nelson Moskowitz. Moskowitz had a bachelor's in physics from Yeshiva University in Manhattan, a master's in physics from Stevens Institute of Technology, and a law degree from Catholic University. He had been with the Patent Office since 1970. Moskowitz, too, said that Gould would get nowhere if he insisted on pressing claims to laser oscillators, and pointed out that the application now before him highlighted amplifiers among its other claims.

Samuel took the argument to Gould, just as Keegan had done. "Amplifiers can work even if oscillators don't," he said. "Let's press for patents on amplifiers, then anybody who builds an oscillator will infringe, because they all have amplifiers."

Gould still objected. He had originated more amplifier schemes than anyone, but he remembered that what felt like the real moment of invention, the big "Eureka!", had been realizing that the parallel mirrors were the key to the laser beam.

"That's mine, and I deserve it," he told Samuel.

"That may be true," Samuel said. "But you lost the oscillator claims, and you have to conduct the legal proceedings accordingly. You can't fight what you lost. That's the reality. You don't have to like it. You don't have to agree with it, but you have to accept it for us to go forward."

Gould grudgingly conceded.

■

Optelecom, as it extended its research into fiber optics, was beginning to encounter problems. Fiber optic cable was all but impossible to buy commercially, and Optelecom had to make its own for its experiments. One of Culver's employees was pulling the fiber in his basement, a process that required hydrogen tanks and hydrogen torches, and the guy had his kids sleeping in the bedroom right upstairs.

Culver told Gould it was time to establish a proper laboratory, and place all of the company's facilities and talent under one roof. It no longer made sense to maintain Gould and his New York Poly cohorts in rented lab space on Long Island.

"I don't see there's any choice," he said. "You're going to have to move to Washington."

Optelecom took space northwest of the city, in Gaithersburg, Maryland. Gould and Marilyn chose Arlington, Virginia, a beltway ride away but closer to Washington's cultural offerings, including the Kennedy Center. Their apartment on Nash Street, just across the Potomac from the District of Columbia, overlooked the Iwo Jima Memorial, and they could see the Washington Monument out their terrace window in the winter when the leaves were off the trees.

35

Samuel and the new team of lawyers proceeded cautiously. Patent Office rules for restricting patent claims require an inventor, when he claims a number of inventions, ultimately to file a separate—or divisional—application for each. Gould's 1974 application revealed several inventions, but Samuel focused first on optically pumped amplifiers.

Here, the Hellwarth Q-switch case that had gone against Gould was his ally. The ruling had opened the door for Gould to claim the amplifier. The Schawlow and Townes patent was still in effect. It described optical pumping of potassium gas by potassium light, but the Patent Office had now implied in the Hellwarth ruling that it wouldn't have worked. Gould's original application preceded anyone else's, and having knocked out Wieder's amplifier claims the field was open.

Samuel worked closely with the new examiner, Moskowitz. He knew that if Moskowitz wasn't going to allow amplifiers, he was not likely to allow any of the other claims. It would be a waste of time and money to file more divisional applications until he knew about amplifiers.

In May of 1977, nine or ten months after Samuel's firm came into the Gould case, Moskowitz sent word to Samuel that he was going to allow the optically pumped amplifier patent.

Samuel and his colleagues could not contain their joy. They shared the news with friends and relatives, several of whom made purchases of Refac stock that would come back to haunt them in insider trading accusations when the over-the-counter stock almost tripled in one week, moving from 2 7/8 to 8 1/4.

■■

Before the patent issued, Samuel prepared and filed several divisional applications. The first, filed on August 11, received Application No. 823,611. It claimed amplifiers using collisions of the second kind in a gas discharge. Although Gould had dropped the interference with Javan and Bell Labs, Keegan's negotiations with the Bell Labs attorneys had narrowed that patent to its particular specifics—an oscillator using helium and neon in a gas discharge to produce a population inversion. As Keegan had hoped, there was much left for Gould to claim in the way of amplifiers. Two months later, on October 6, Samuel filed claims on the uses of lasers that Gould had so accurately predicted, including the heating of materials, distance measuring, communications systems, television, and certain industrial applications. This was Application No. 840,050. And on the same date he filed Application No. 823,665, claiming a "polarizing apparatus employing an optical element inclined at Brewster's angle"—the use of Brewster's angle windows in a laser.

Filing the divisional applications before the amplifier patent issued would allow Gould's applications to receive the benefit of co-pendency with his others, and thus the benefit of the 1959 filing date.

Meanwhile, Gene Lang at Refac had a patent to license at last. His first letters to laser manufacturers let them know of the pending Gould patent and the claims it covered. Then, in registered letters dated October 4, 1977, he made his demands.

The letter he wrote to Mel Cook at Holobeam Laser, Gould's former employer, mirrored the others: "We are pleased to advise that U.S. Patent 4,053,845 will issue to Gordon Gould next Tuesday, October 11.

"Accordingly, beginning with October 11, we will regard all lasers that employ optical pumping amplification subject to license under the Gould patent. This would include lasers manufactured by your company." He went on to offer a license, "together with possible license options on future laser patents of Gordon Gould, at a royalty rate of 3.5% of the net selling price of lasers covered by the patent."

Lang tacked on a history lesson. "You are probably aware that the Gould patent claims were allowed . . . only after a most exhaustive examination and with due regard to decision of all interference actions." He offered to supply the Patent Office's brief supporting its decision to allow the patent, and Refac's legal analysis of Gould's patent position. It was all to say that there was substantial evidence supporting the Gould patent, and that fighting it was useless.

He gave the manufacturers a ninety-day window in which to take the deal.

■■

U.S. Patent No. 4,053,845 issued as expected. It gave Gould the credit for the kind of amplifier that had excited the first laser, Maiman's ruby laser. It was almost twenty years since Gould first wrote down his conception of the laser, and more than eighteen years since he filed his patent application.

He put on a light-colored V-neck sweater and a tie, and posed at Optelecom's laboratory for a photographer from *The New York Times*. The picture that appeared later that week showed a smiling Gould, eyes lively behind his thick, dark-rimmed glasses. Gould had turned fifty-seven the previous summer and the years showed, too, in the pouches under his eyes and the lines descending from his nose to the corners of his mouth. The story that accompanied the photo referred, with some understatement, to the "long history of litigated oppositions and interference actions." It also accounted for Gould's smile, with his estimate of the world market for the amplifiers at between $100 million and $200 million annually, "promising him substantial royalities for seventeen years," the life of the patent.

A second week-long buying spree pushed Refac's stock from 9 1/2 to 15 1/5.

The laser industry did not share in the euphoria.

Dick Samuel watched all hell break loose. In all his years as a patent attorney, he had never seen a big reaction when a patent issued. This time, however, the laser makers went insane. "It was like the world was coming to an end," he said.

The industry had continued to expand at a rapid pace. In 1977, estimates placed annual sales at $400 million. Manufacturers had been paying 2 percent royalties to the Research Corporation and Charles Townes on his maser patent since 1971, after Spectra Physics settled out of court rather than risk a trial against the Nobel laureate, and other makers had fallen into line. Townes's patent had expired just the year before. Gould's patent covered about a third of all lasers made, and Lang had made it clear he expected to be paid.

"We're not going to grab off more than we're entitled to," Lang told *Business Week* magazine. "But we are going to insist on our due."

Herbert M. Dwight, Jr., the president of Spectra Physics, complained in the same *Business Week* article about the prospect of paying royalties again. "The whole thing stinks so far as I'm concerned," he said.

Control Laser, Inc., of Orlando, Florida, ranked third or fourth among laser manufacturers. Its president, M. Lee McDaniel, vowed to fight the Gould patent. He told *Business Week*, "We will combine with all interested parties, and fight this thing as hard as we can."

It was a bold statement, and a risky one. While Lang at Refac had been urging licenses on the manufacturers, Samuel had been working on infringement lawsuits in anticipation that they would resist. He read the article, and decided that McDaniel had nominated Control Laser to be sued. The ink was hardly dry on the article—*Business Week* had a cover date of October 24, but appeared a week earlier—when he filed against the company in federal district court in Orlando on October 19.

36

Far from the commercially driven world of laser manufacturing, in the academic halls where prestige and recognition were more valuable coinage than patent royalties, others also looked with dismay on the Gould patent.

Lang and Refac had prepared a triumphant eleven-page news release that coincided with the patent grant. With the optically pumped amplifier patent, and five others pending, it said Gould's footprint covered "virtually the entire laser industry."

"Gould's position in this respect, as 'father of the laser,' in no way denigrates the enormous value, consequence and contribution of many other brilliant physicists such as Townes, Schawlow, Maiman, Hellwarth," crowed the release. "By the same token, the work of all these eminent scientists cannot detract from Gordon Gould's rightful position which, after eighteen years, is coming to fruition."

Gould, dismissed as an "attic inventor" after losing the Schawlow and Townes interference more than ten years earlier, was rising like the phoenix from the ashes. Here was *Business Week* saying that Gould's big break came in the 1973 decision in Hellwarth that, as the magazine put it, "the Schawlow-Townes patent did not, after all, contain enough information to tell anyone how to make certain key parts—particularly the amplifier—of a laser." Saying the decision "invalidated much of the Schawlow-Townes patent." Referring to "the holes in the Schawlow-Townes patent." And *The New York Times,* giving credence to Gould's claim of inventing the very term "laser" by quoting the heading from his 1957 notebook.

The "attic inventor" was reemerging as a laser pioneer, at least in the public eye, while Townes's claim to the laser's underpinning theory, and his and Schawlow's patent, were being superseded.

This did not sit well with Townes. He had, since the 1973 decision, second-guessed the Bell Labs lawyers who had written and filed his and Schawlow's patent application. Now, once again, he revisited the events of July 1958 and remembered his annoyance at the apparent reluctance of the lawyers to seek a patent.

"I never paid much attention to the patent," he said. "The result was, it was not really very well written. There were a number of things in our paper which didn't get carefully covered."

It was the lawyers' fault, in other words.

Schawlow, interviewed by *Business Week,* professed to being mystified by the patent process. "Patent law is a funny thing," he said.

Gould agreed, but not unhappily. Maybe Maiman should have received the patent, he said. Then again, he still thought he should have been given the patent for the Q-switch. "It seems random," he said.

Townes, on the other hand, began to say when he was asked not only that Bell Labs' patent lawyers had overlooked claims contained in his and Schawlow's paper that allowed Gould "to sort of pick off various things that we had in our publication paper but weren't mentioned in the patent." He would say that Gould had seen the paper in a way that implied that the ideas in Gould's grant proposal to ARPA and in his patent application had really come from Townes and Schawlow.

As time went on, he would hint that there was something funny about the fact that the notary who had put his name to Gould's 1957 notebook also was named Gould. He would hint that the notebook might have been misdated. He verged on accusing Gould of outright theft, and he acted ungraciously.

And Townes's attitude was typical within the scientific community. The politics of science elevated Townes, who by late 1977 had added still more lectureships, honorary degrees, awards, and committee and board memberships to his resumé, while the keepers of the grail denigrated Gould as a product of the patent lawyers.

■■

The laser industry's outrage was pervasive. Resistance mounted, bringing a quick chill to Gould's euphoria. On November 23, 1977, six weeks after the Patent Office issued the '845 patent, examiner Moskowitz filed a First Official

Action in which he rejected Gould's application for the gas discharge laser amplifier using collisions of the second kind.

He stated several reasons, but the most curious were two that seemed to contradict each other. He held, as had the Board of Patent Interferences and the Court of Customs and Patent Appeals in Gould's interference with Hellwarth, that Gould had not adequately disclosed the invention he claimed. The court had ruled with regard to oscillators, but Moskowitz said its ruling had included amplifiers. He also rejected Gould's application on grounds of obviousness—that is, the invention was obvious to persons "skilled in the art," and therefore unpatentable.

If an invention is insufficiently disclosed, so that somebody who ought to know how could not build it, it could hardly be obvious to the same person.

Samuel, shaking his head in wonder, undertook a host of amendments to the application to try to deal with Moskowitz's objections. He focused it more directly on amplification by collisions, and deleted material from the original application that applied to the already-patented optically pumped amplifiers. The negotiations extended into 1978.

Meanwhile, the laser manufacturers were joining forces to avoid paying Gould royalties.

They retained a San Francisco law firm, McCutchen, Doyle, Brown and Enerson, with offices in the Embarcadero Center. Soon, in January 1978, a draft "Defense Fund Agreement" went out to nine laser makers in addition to Control Laser. The fund's aim, according to the agreement, was "to insure the availability of funds sufficient for a proper defense of the case of Gordon Gould v. Control Laser." The law firm and its counsel, Terry J. Houlihan, would administer the fund, and advise its members on legal aspects of the fund.

Control Laser would pay 25 percent of the cost of defending itself against the infringement suit. The remaining three-quarters, according to the draft agreement, would be divided among the participating manufacturers according to their percentage of the total sales of commercial optically pumped lasers in the previous year. Military laser sales were exempted from the calculations. The proposed agreement set out billing and accounting procedures, and stipulated that the agreement would terminate when the last legal bill was paid, or when the number of participants dropped to five or less.

The draft went, in addition to Control Laser, to GTE Sylvania, Raytheon, Apollo Lasers, Coherent, Korad, General Photonics, National Research Group, Quantronix, and Spectra Physics. Quantronix, of Smithtown, New York, on Long Island, was headed by Dr. Richard T. Daly, Gould's old nemesis at TRG.

January was not over before Samuel had obtained a copy of the draft agreement, demanding it in the course of deposing Control Laser chairman R. D. "Robby" van Roijen in Orlando. Houlihan blasted van Roijen and Orlando attorney Robert W. Duckworth, who was defending Control Laser against the Gould suit, for turning over the agreement. He warned the fund's other potential participants about turning over information to the enemy, and said the defense fund was "no more than a financing mechanism" and wasn't relevant to either liability or damages in a lawsuit.

Samuel thought otherwise. He amended his complaint against Control Laser at the end of February, accusing the company of antitrust activities. He said Control Laser had instigated an industry-wide boycott of Gould's patent unless Gould agreed to accept the industry's royalty fee scale. They are "conspiring with each other to continue to infringe, induce each other to infringe, and interfere with the plaintiff's legal rights," Samuel charged.

His countertactic was to sue each manufacturer separately. The first action came against General Photonics, a Santa Clara, California, manufacturer charged with patent infringement and antitrust activity in a suit filed in federal court in San Francisco. General Photonics was not a giant of the industry. With only $10,000 profit on sales of about $825,000 worth of lasers in 1977, it was in no financial shape for an expensive legal battle. Nor was its lawyer confident about going toe-to-toe with the aggressive litigators Gould had in Samuel and his colleagues. The lawyer, James R. Hagan, wrote Duckworth in August 1978 asking for assistance.

Duckworth passed the request on to van Roijen. Control Laser's chairman responded, rejecting Hagan's request but pressing again for cost sharing among the manufacturers. General Photonics would be on its own.

By then two more firms, Molectron and Quantra-Ray, had been added to the defense fund mailing list.

■■

While the laser industry was spreading its legal costs around, Gould's financial well was running dry. The $300,000 worth of legal time that Samuel's law firm had committed in exchange for a share of the patent income—an amount that at the outset had seemed virtually unlimited—went quickly, pursuing the patents and pressing infringement suits in two jurisdictions a continent apart.

The industry's tactics could not have been more effective. Not only was an entire industry distributing the burden of legal costs, its resistance was forcing the other side to exhaust its own financial resources, not only to collect royalties, but to secure Gould's other patents.

Moskowitz's resistance to Gould's collisions amplifier patent was as stubborn as his willingness to grant the optically pumped patent had been cooperative. Samuel sensed some resistance that went deeper than the examiner's research on the history of the patent. Moskowitz seemed to be subjecting the application to an unusual level of scrutiny.

Samuel believed that the industry outcry had been so swift and severe that it had spooked the Patent Office. "There was a thing going on in the PTO," he said, using the initials for the Patent and Trademark Office, "not to allow any more Gould patents. There was pressure from the industry, tremendous pressure."

Whatever lay behind it, the rejection meant more legal costs as Samuel tried to conform the application to meet Moskowitz's objections. But on May 19, 1978, the examiner issued his second and final rejection of the patent, still citing obviousness and insufficient disclosure in the same breath. He dredged up the Hellwarth case again, and repeated his insistence that the ruling against Gould had included amplifiers.

Gould was headed back to court. Facing yet another level of appeal in yet another patent quest, Gould's team now faced more costly depositions and filings. Gould had bartered away 65 percent of the value of his patents. He had one patent to show for it, and dwindling funds to try to enforce it.

When the $300,000 war chest was exhausted before the end of 1978, Dick Samuel knew that the deep pockets of the industry would prevail if he could find no new source of funding.

37

Panelrama stores dotted the landscapes of five mid-Atlantic states. The small chain of thirty-five stores sold pre-finished plywood paneling and other home improvement supplies and hardware to do-it-yourselfers in Pennsylvania, New Jersey, Delaware, Maryland, and Virginia. It was headquartered in the Philadelphia suburb of Ardmore, Pennsylvania.

Panelrama's founder and chairman was an entrepreneur in his mid-thirties named Gary Erlbaum. "Motivated by poverty," as he liked to put it, Erlbaum had opened his first Panelrama store in 1965, when he was still an undergraduate studying business at Temple University in Philadelphia. He went on to law school, and took Panelrama public in 1972 with the advice of a New York investment banker named Kenneth G. Langone, who had engineered the wildly successful public offering of shares in Ross Perot's Electronic Data Services in 1968. The chain had had its ups and downs. Erlbaum had wrestled it back to profitability but now, early in 1979, had decided it was just too difficult to run. Sensing that his future lay elsewhere, Erlbaum turned again to Ken Langone.

Erlbaum told Langone he thought he had three choices. He could buy another company, sell out to someone bigger, or go in a new direction. He was open to suggestion. He and Langone swapped ideas, and they agreed to meet in New York the following week. Erlbaum drove from Philadelphia for the dinner meeting with his younger brother, Steven. They joined Langone at his office in the Seagram Building. Later, as they walked along Park Avenue on their way to a restaurant, Langone said he thought he had an answer to Erlbaum's problem.

The story Langone told, in a voice hoarse enough to have been sanded with rough paper, was the kind of story that fuels a thousand dreams on Wall Street and in the investment banking houses every day. This one had come to Langone by way of an unsolicited phone call from the Westfield law firm. They had been referred to him by a vice president at Shearson, who had met with the lawyers and heard their story.

"My guys won't do it," she told them. "But you should call this guy, he's known as the Wild Wop of Wall Street."

Langone had been intrigued by what he heard on the phone, he told Erlbaum and his brother. He had talked with Dick Samuel, and later met with him and Gordon Gould. Langone, the son of a Long Island plumber and a high school cafeteria worker, had not achieved his corner office overlooking Park Avenue by misjudging character. Tall, balding, and profane, Langone had amassed a fortune by betting on "whether people are bullshitting me or telling me the truth." When Gould told him he had been fighting for years for the laser patents to which he was entitled, Langone thought the statement was preposterous. "Well, that's interesting," he said, "because I've always thought that I should have the patent on the wheel."

Gould had let the insult pass. People who heard his story were always incredulous at first. He didn't really blame them. He had been through far too much to be fazed by Langone's reaction, and so he just kept talking.

Soon Langone found himself nodding in response to Gould. The son-of-a-bitch had a far-fetched position, but he sure believed what he said. Everything, from his body language to his calm and detailed responses to Langone's questions, told Langone Gould was on the level.

The party had reached the restaurant now and been shown to a table. Langone looked across the table at Erlbaum and said, "Gary, I don't know if this guy's smoking dope or what, but he sure is convinced. This may be the ideal thing."

Gould's story appealed to Erlbaum. A passionate tennis player, he left blood on the court each time he played. Causes and issues also could incite his passions, and nothing appealed to his emotional side more than a valiant underdog. As Langone described Gould's long battle, Erlbaum began to see not only business opportunity, but a chance to aid the cause of justice.

With Erlbaum's okay, Langone set up a meeting with the lawyers from Westfield.

■■

The meeting, held in late June of 1979 at the law firm's offices, was filled with talk of glowing prospects, especially on the lawyers' part. Their optimism verging on hubris seemed justified. Two months earlier, in April, the Patent Office had sent word it was going to issue Gould a patent on uses of lasers. Like the optically pumped amplifier patent, this was a basic patent, only it went even further, covering most of the lasers used in industry and in such applications as photocopying machines.

The lawyers, primed to expect across-the-board resistance from the Patent Office, weren't expecting such a glorious piece of news and they shot off their mouths. Samuel and Littenberg got duded up in vested suits—Samuel even wore a watch chain—for a photo shoot for *Business Week*, and Samuel told the writer of the accompanying article, "Gordon Gould invented the laser, but we invented the laser patent application." Somehow the groundwork laid by Keegan based on Wibert's advice had been forgotten.

Gould was quoted, too, and he was both more sensitive and more accurate. He said he was happy to have the patent, but "I'm afraid it's going to shake up the industry and cause a lot of consternation."

Samuel and company still were feeling pretty good by the time the meeting with Langone and Erlbaum rolled around. They said they saw the Patent Office's resistance to Gould's gas discharge amplifier patent application as a temporary glitch that would easily be overcome, once they had the resources to litigate.

Langone told Erlbaum, "This thing has been going on for a long time, these lawyers are good, this thing should be resolved in a year."

In short order, they had the makings of a deal. The parties signed a letter of intent under which Erlbaum would liquidate Panelrama's assets and merge them with Creative Patents, Inc., into a new firm created for the purpose of securing the Gould patents. The law firm's telex address was Patlex, a combination of "patent," and the Latin word for law, and that became the name of the new company.

In exchange for the war chest that Panelrama's liquidation would produce, Patlex would own 40 percent of the future income from Gould's patents—the 25 percent already owned by Samuel's law firm, plus an additional 15 percent Gould agreed to sell the firm for $2.3 million. Gould was now down to a 20 percent share of his invention, while Refac's 40 percent share would remain unchanged.

Before he signed the papers, Erlbaum took one last step. He hired a law firm in Philadelphia, with no ties to either side, to assess the deal. The report came back with the positive spin Erlbaum had hoped for. It said that Gould was going to win.

■■

The Patent and Trademark Office issued the use patent, No. 4,161,436, on July 17, Gould's fifty-ninth birthday.

It had not come from Moskowitz, Samuel had learned, but from an examiner named Howard S. Williams. Samuel could only surmise that Williams was out of the loop at the Patent Office. He apparently hadn't heard that Gould's patent applications were to be looked at with extra scrutiny.

"He just issued it, oblivious to the world," Samuel said. "It was weird, like he wasn't part of the team. Somehow he got this application, and no one ever told him."

Three types of uses were covered in the patent. Dissociating, or vaporizing, a material covered welding, drilling, and cutting of anything from plastic to steel. Promoting a chemical reaction covered the laser-based photocopying machines that were coming into widespread use. Stimulating the release of energy would cover laser fusion, if it ever got beyond the research stage.

The '436 patent gave Gould important new sources of potential royalties. Twenty-two years after he had foreseen the heat that concentrated laser light could bring to bear, lasers were finding ever newer applications that included photochemistry, laser fusion, and the separation of isotopes by laser. They already were in wide use in manufacturing. They had become an indispensable tool in automaking, for example, where they cut and welded metal, measured, and aligned work with precision.

At Refac, Eugene Lang signaled that he would ask for a royalty based on the total value of the product that the laser made possible. He had done something similar with the optically pumped amplifier, asking for a percentage of the total laser system.

As before, the laser industry dug in its heels, contending the patent wasn't valid and preferring to contest it rather than negotiate licensing fees.

Now it made sense to go after laser users as well as laser makers for infringement. As a first step, Samuel filed infringement suits in Chicago's U.S. District Court, and in Canada's federal court, against Lumonics, a laser maker in Kanata, Ontario, Canada. Lumonics was Canada's biggest laser maker. Its lasers were used by General Motors on its assembly lines in the United States and Canada.

Lumonics countersued in Chicago, challenging the validity of the U.S. use patent. The issues in Canada were different, because there Gould had been issued a more far-reaching patent that covered the laser *en toto.* Gould *was* the inventor of the laser, as far as Canada's patent office was concerned.

Samuel had expected Lumonics to respond. He hoped, however, that the laser maker would be his sole adversary. He assumed that General Motors also was assessing Gould's suit and considering its options. The automobile giant, like other manufacturers whose assembly processes incorporated lasers, had licensed the now-expired Townes maser patent and looked dimly on the prospect of paying new laser royalties. GM used $10 million worth of lasers in nineteen different manufacturing processes. Along with the laser makers, it had a stake in killing Gould's patent.

When GM decided to join Lumonics as both a co-defendant and co-plaintiff against Gould, Samuel knew that the use patent had not only validated Gould's earliest vision of the laser, but had also acquired him a new and powerful set of enemies.

■■

The deal that created Patlex closed on December 11, 1979, Erlbaum's thirty-fifth birthday. Going-out-of-business signs went up in Panelrama stores. Fixer-uppers carted home Panelrama's inventory at bargain prices, and fed the coffers of a landmark patent litigation. When nothing was left but the walls, Erlbaum sold off the stores, and the property on which they sat. He netted close to $2 million.

Gould received a $300,000 down payment on the additional 15 percent of future income he was giving up, and a seat on the Patlex board. The rest was due in eighteen months, when Ken Langone expected to broker a public offering of additional Patlex shares. Gould's patents would surely be in place by then, and the income they produced streaming into the Patlex treasury. Dick Samuel, spirits buoyed by the use patent and his rejuvenated litigation fund, looked enthusiastically to the year ahead.

38

Early in 1980, a long-sought revision of U.S. patent law finally was at hand. Representative Robert W. Kastenmeier of Wisconsin introduced House Resolution 6933 on March 26. The proposal aimed to add to the patent laws a provision for the Patent Office to reexamine patents at the request of outside parties who produced prior art, or publications, that the Patent Office had overlooked in issuing those patents.

Prior art is a sacred component of the patent law. Inventors, when they submit their patent applications, are required to provide a list of references. This helps examiners determine whether the invention is really new or not, and assures that applicants aren't taking credit for something that somebody else already thought of.

The change in the law was, in theory, a good idea. Individual inventors and small companies had long advocated an administrative procedure that would let them avoid costly court battles over patents. As it was, the courts were the only route for challenging an issued patent. Patents could be reexamined by the Patent Office, but only at the request of the patent owner. That meant an inventor with an earlier claim had no choice but to go to court. Patent owners often could continue to assert invalid claims because challengers lacked the money or the will to undertake litigation. As early as 1966, the President's Commission on the Patent System recommended a reexamination procedure for cancellation of claims, "which should be faster and less costly than court proceedings."

The commission recognized that the process could be used to harass patent holders. It recommended that high fees accompany requests for reexamination, and that anyone who sought cancellation unsuccessfully be required to pay the patent holder's defense costs.

The proposal for third-party reexamination requests languished for years. Once introduced, however, the bill rode a fast track. Barely three weeks after Kastenmeier's bill entered the legislative machinery, the House Judiciary Committee's Subcommittee on Courts, Civil Liberties, and the Administration of Justice convened hearings. Donald R. Dunner, president of the American Patent Law Association, spoke to the subcommittee on April 15, 1980. He said the association, which represented patent lawyers and inventors—it is now the American Intellectual Property Law Association—"wholeheartedly supports legislation providing for reexamination of issued patents by the Patent and Trademark Office."

Dunner cited out-of-control patent litigation costs "conventionally reaching levels of $250,000 or more. While such a substantial financial burden can be borne by large corporate patent owners, it makes patent enforcement or defense an almost prohibitive undertaking for individuals and many small businesses."

He went on to say that the Patent Office should be given the resources to examine patents more effectively in the first place. But the reexamination proposal, he said, was the best means of considering inventions and publications that the first examination failed to take into account.

Reexamination's advocates, including Dunner, saw its potential for harassment of patent holders. Dunner, however, was mollified by "a built-in harassment-prevention device." That was the provision that the Patent Office would take a new look at a patent only when the Patent Commissioner determined that the third party requesting reexamination had actually raised "a substantial new question of patentability." Third-party involvement was somewhat limited under the proposed law. This, Dunner said, "comes close to simulating the ideal procedure." That would have been the Patent Office's finding the applicable prior art and judging the patent correctly in the first place.

The resolution sailed through the hearing process with minor amendments. The House passed it by voice vote on November 17, 1980, and sent it to a Senate eager to get home for the Thanksgiving recess. Both houses agreed on amendments on November 21, and sent the legislation to the White House. President Jimmy Carter signed it on December 12, and Public Law 96-517 took force on July 1, 1981.

Corporations unhappy at the prospect of paying royalties to Gould took notice. At Refac, Eugene Lang told a colleague, "I'm beginning to think that God's against us."

∎

Optelecom, started on Bill Culver's idea for laser-based missile guidance, was evolving into a developer of optical cables and laser illuminators while it continued to work on laser guidance systems. Culver, after Rand and the Institute for Defense Analysis, knew Washington and had turned Optelecom into a typical "Beltway Bandit" that depended on government contracts as its financial lifeblood. It did work for ARPA, and the Army's Advanced Concept Teams I and II under the Army's Office of Chief Scientist. The ACT contracts were monitored by the Army Missile Command from its Redstone Arsenal headquarters in Huntsville.

Culver was Mr. Outside in the scheme of things. He did the selling, and Gould did the inventing. Gould also oversaw all of the laboratory work, and managed some of the research programs.

Fiber optics, begun as a means of carrying light and images over cables with low reflective losses, also was evolving, into an information medium. Nobody knew much about the way the cables carrying threads of light-borne information would be used, or even how to make them. The new objective of Gould's inventiveness became finding ways to combine the new field with the information needs of various industries and agencies.

Optelecom got a contract from Chevron to develop an oil well logging cable. Traditional wire logging cables didn't have the bandwidth to deliver the information drillers needed to the surface. Gould, working with several subcontractors, devised a cable combining three optical fibers. At the business end, deep in a hole in the ground, neutron pulse generators bombarded subsurface material. The neutrons were absorbed by atoms in the surrounding rock, which in turn became radioactive and gave off signals that allowed them to be identified. The cable's fifty-four elements were wound with such precision that the fiber optics maintained perfect coherence for the length of the cable, and delivered a wealth of information to the surface. The signals told the drillers whether they were likely to find oil.

The cable was expensive, and as oil companies scrimped on research funds a second was never produced. The lone cable remained in use for over fifteen years.

Gould's tenure at Optelecom would produce five patents. None, however,

were of the importance of the basic laser patents. Culver and Gould, although they were the principal owners of the company, each made $49,000 a year. It was a reasonable living, but not a lavish one. Culver saw in Gould an increasingly urgent recognition that his laser patents offered the only escape from the middle-class treadmill.

Gould's focus on the patents was buttressed by a certain recklessness. He had celebrated his sixtieth birthday in the summer of 1980, but he was still smoking three packs of cigarettes a day and enjoying a cocktail, or several, in the evenings.

Culver had started to keep an eye on Gould, and thought he was starting to show wear. He believed his partner drank and smoked too much, an opinion that was reinforced on mornings when Gould couldn't keep the hand that held his coffee cup from shaking. He knew Gould's doctor had told him to quit, or at least cut down, on his twin vices. Finally, Culver bluntly told Gould he should stop. Gould paid no attention.

His debilitating lifestyle aside, Gould actually was living better than he ever had. The $300,000 he had received at the consummation of the Patlex deal had allowed him, for the first time in his adult life, to move out of an apartment. He and Marilyn had left the building on Nash Street in Arlington in the spring of 1980 for a house on two acres in the outlying Washington suburb of Great Falls, Virginia, several miles up the Potomac. The seller had agreed to hold a 15 percent mortgage, a financial boon in those days of 20 percent interest rates. Marilyn had a job she loved, working at George Washington University's medical school, where she used computer analysis of twenty years' worth of electrocardiograms to develop statistical predictors for heart attacks.

More than in any of Gould's previous relationships, he shared with Marilyn a mutual respect and sense of partnership. Her perceptions and personality complemented his in many ways. In fact, she completed him; her belief in him was total, and her hard-edged, real-world perception, honed at NYU law school, was an antidote to the vagueness Gould sometimes brought to the negotiating table when it came to business. Marilyn was good for Gould, and after too many relationships that had ranged from disastrous to merely ill-advised, he had sense enough to know it.

■■

Patlex had eighteen months from its formation to complete the purchase of Gould's additional interest. In the late spring of 1981, the deadline was approaching and Ken Langone was looking with alarm at the deal he'd put together. The euphoria that had attended Patlex's organizational meetings had faded. It was clear that it was going to take long and costly litigation to enforce Gould's two existing laser patents. The two others remained pending. The gas discharge amplifier patent would cover a large portion of the laser industry if it ever was granted, but the Patent Office clearly was going to continue to resist.

Langone saw himself going into the marketplace offering shares that nobody—so far—had a compelling reason to buy. Afraid the offering would fall flat, the investment banker told Gould he wanted an extension.

Now it was Gould's turn to dig in his heels. He had money coming for which, from his perspective, he had waited a long time already and he told Langone in a phone call that if Patlex wanted an extension, the company would have to pay. He had signed away 80 percent of the laser patents' proceeds, and he was tired of living—albeit comfortably—from paycheck to paycheck.

Langone was apoplectic. Shouting over the long distance phone line from New York to Virginia, he demanded that Gould resign from the Patlex board. Gould faxed his resignation immediately to Gary Erlbaum, who went apoplectic in turn. Gould's board membership gave Patlex the scientific credibility it badly needed to appeal to investors. In a furious series of phone calls involving Gould, Erlbaum, Langone, and too many lawyers to easily count, Gould was persuaded to rescind his resignation.

Then, with the public offering pending, he and Marilyn boarded an Air France flight from Dulles Airport to Paris. Once in France, they disappeared from the Patlex radar screen. They spent two weeks floating along barge canals in southern France. Their barge's many luxuries included the fine wines of Bordeaux and Burgundy and exquisite meals placed before them three times daily. But to Gould's mind, and Marilyn's, the greatest luxury of all was the absence of a telephone.

They returned from France in late June and flew into New York. Not knowing whether the offering had been successful, they had booked a room at the luxurious Plaza Hotel. Marilyn told Gould, "Either we can't afford this, in which case we'll have a last fling, or we can, and it won't matter."

They could afford it. Investors had snapped up the available shares in the Patlex offering. Gould's $2 million balance had been wired into his account.

Later, when Gould decided that he wanted to share some of his patent wealth with his relatives, with Marilyn, and also with Tanya Bogart, with whom he remained friendly, he formed a corporation to parcel out 19 percent of the shares. But he had been perplexed about what to call it. The creativity he found easy to muster when it came to inventing abandoned him. Eventually, he settled for NGN Acquisitions—the initials stood for No Good Name.

∎∎

Around October of each year, when the Nobel Prize rumor mill started to hum and sing, Gould found himself paying attention. He liked to see which of the old boys—most of whom he knew or had had some contact with—his colleagues had chosen to drape with the mantle of greatness. With his $2 million in the bank, Gould told himself he had the one reward he really cared about, but the truth was he had always felt a little slighted when the awards were handed out.

He had always said he didn't want to play the game, but something in him still wanted to be asked.

So he reacted with some interest when Arthur Schawlow was announced as one of the Nobel recipients in physics in the fall of 1981. Expecting the worst, he scoured the news reports to see if the Swedish Royal Academy had named Schawlow belatedly as an inventor of the laser.

The academy, in fact, had said nothing about inventing the laser in Schawlow's citation. He and Nicolaas Bloembergen of the United States shared half of the prize for their respective contributions to laser spectroscopy. Their work, and that of the other winner, Kai M. Siegbahn of Sweden in electron spectroscopy, had "made it possible to investigate the interior of atoms, molecules, and solids in greater detail than was previously possible," said Ingvar Lindgren of the academy in awarding the prize. It was a classic Nobel award, in that it stressed science rather than widespread application.

That he shared the prize with two other physicists did not diminish the affable Schawlow's sense that he was done at last with a long journey. The expectations that had been awakened seventeen years earlier, when the Stanford University publicity department took his picture in mistaken anticipation, finally were put to rest.

By then Schawlow was known around Stanford and the San Francisco Bay area not just as a physicist, but also as a devoted fan and collector of traditional jazz. He played jazz clarinet, and had amassed an encyclopedic knowledge of the form, along with thousand of records, many rare and valuable. He approached the music with a pleasure as great as that he brought to teaching and research. Teaching to Schawlow meant reaching out through television and other media to demystify science for lay people. Schawlow's scientific colleagues generally agreed that the Nobel Prize couldn't have have happened to anyone nicer. They also knew that Schawlow had maintained his equanimity in the face of a long-running family tragedy; he and his wife, Aurelia, had three children, one of them a son who was autistic. He was now an adult who needed full-time care. The Schawlows had devoted much of their lives to establishing appropriate residential care for sufferers of autism in California, and they visited their son each Friday, a trip often requiring long drives.

There were those who said that Townes had lobbied hard behind the scenes to get his brother-in-law the coveted Nobel. Townes, like all Nobel laureates, was among those asked annually for nominations, and he routinely nominated candidates in both chemistry and physics. No doubt he talked up Schawlow for the prize. But Schawlow's work also had deserving scientific weight.

Schawlow saw an advantage in his long wait for the Nobel. "I'm just as glad I didn't get it earlier," he said. "It was kind of nice to get it on my own and not just as Charlie's assistant. Charlie is a very brilliant man. I'm not anywhere near as smart as he is, but it was nice that I could do it on my own."

39

Gould's patent infringement suit against the Santa Clara, California, laser maker, General Photonics, filed in 1978, was ready to go to trial early in 1982. The company had failed to find backing in the laser industry for its defense. It wanted to stay out of court, and had corresponded with Refac about licensing the optically pumped amplifier patent, albeit on terms it considered favorable. But it was too little, too late. Samuel wanted a court victory to impress the industry, and he smelled one.

The non-jury trial opened on February 8, 1982, in the U.S. District Court in San Francisco. Judge Samuel Conti had just finished trying a Hell's Angels case; the courtroom had bulletproof glass and the chairs, which were bolted to the floor, were outfitted with arm and leg restraints. Dick Samuel caused quite a dust-up, too. He strode into the courtroom and in his opening remarks reignited the controversy that had simmered between Gould and Charles Townes for a quarter of a century.

Gould had been "discredited and abused for twenty years" by a scientific community that insisted he had copied Townes's ideas, Samuel said. But nothing, he added, could be further from the truth. "If anything, Your Honor," he told Judge Conti, "at the end of this trial you may well find that certain of the subject matter which Charles Townes disclosed in his patents were, in fact, written down first by Gordon Gould, witnessed by Townes, and then put into the Townes patent."

Gould, Samuel said, "is the true father of the laser industry today."

The *San Francisco Chronicle* ran a story the next day. The article pulled from trial documents Gould's claims on inventing laser amplifying systems and coining the word "laser." It referred to Gould's having showed Townes "his preliminary work," and the charge that "Townes later obtained patents and was awarded the Nobel Prize."

The implication was strong that Townes had appropriated Gould's work.

Townes, contacted by the *Chronicle's* reporter, shot back that Gould "was aware of what I was doing, and my work was earlier." The great physicist was clearly rankled. His interests had turned from spectroscopy to astrophysics during his tenure at Berkeley, but he jealously guarded the pedestal on which he stood as co-inventor of the laser.

The pedestal shook short weeks later, when Judge Conti ruled in Gould's favor. Every patent infringement case rests on two questions: Is the patent valid, and if so, is it being infringed? The judge said yes in both cases.

The lynchpin decision in the Hellwarth interference, where Gould had lost his claim to the Q-switch but gained the vastly more important opinion that neither he nor Townes and Schawlow had invented an operable laser, echoed in Conti's March 1 ruling. Judge Conti said Gould was entitled to a date of conception as early as August 28, 1958—when the first pages of his second notebook were signed at TRG—and to his original filing date of April 6, 1959. Thus, he ruled, the Patent Office had been right to allow Gould's amplifier claims as a patentable invention over the claims in the Schawlow and Townes 1960 patent. Gould's 1977 optically pumped amplifier patent was valid.

The patent had withstood its first court test. Optically pumped amplifiers were used in perhaps 35 percent of a laser market that in 1982 totaled, by some estimates, $1 billion. Gene Lang at Refac proclaimed the victory would loosen the purse strings of the industry and bring "substantially more" income. Gary Erlbaum spoke of buoyed expectations and desires. Over-the-counter trading in both Refac's and Patlex's shares had been suspended pending announcement of the trial result; otherwise, there would have been a buying frenzy.

The laser industry, however, shrugged off the decision. General Photonics said it hadn't had the money to defend itself adequately. Indeed, Judge Conti had been so frustrated by the Photonics' lawyer's cross-examination of Gould that he questioned Gould himself. General Photonics president Burton Bernard, when the ruling came down, argued no contest. He said he still doubted the validity of the Gould patent, but said he couldn't afford $200,000 to hire a patent attorney to prove it. Much of the firm's case had been aimed at fending off the triple damages that would have resulted from a ruling of willful infringement.

As it was, Gould and General Photonics had agreed in advance on a settlement if Gould won. General Photonics agreed to pay $75,000 in back damages, and 8 percent of future sales in royalties. But even that was too great a burden for the firm, which had about twenty employees. Anticipating Conti's ruling, General Photonics had agreed two days after the trial to be acquired by Hadron, another laser maker, in a stock swap.

Hadron's acquisition demonstrated the many clonings, spinoffs, and permutations of the still-growing laser industry. Now located in Virginia, Hadron was the same firm that had acquired TRG's laser operations from Control Data in 1970.

■■

Townes and Schawlow, like the laser industry, seized on the idea that Gould would not have won an all-out legal fight. "This case wasn't defended at all," Schawlow told *The Wall Street Journal,* his legendary affability crumbling when it came to Gould. "I have grave doubts that this [Gould] patent will be found valid when it is properly argued."

He was hotter still a few days later, when he was interviewed by the *Chronicle of Higher Education* for an article that appeared on March 17. "This man is getting away with murder," he told the *Chronicle's* reporter. "What he patented was clearly foreshadowed in our patent."

Townes, interviewed for the same article, took a typically dismissive tone. Gould's amplifier patent was merely a "third-tier patent" on an aspect of the laser his—Bell Labs'—patent lawyers had told him was unpatentable because it occurs in nature. Once again, Townes was blaming Bell Labs' patent lawyers for allowing Gould into the game. As for the Patent Office's awarding Gould's patent, Townes said, "The Patent Office frequently is wrong."

The brothers-in-law both struck out at the implication that Townes had lifted optical pumping ideas from Gould. Schawlow called the idea "a ridiculous canard." Townes told the *Chronicle* much the same thing. "Any claim that I took his ideas is ridiculous," he said.

Townes used the interview to draw a line in the sand. He said Gould's claims were "easily disprovable." Townes had not testified in the San Francisco trial, since General Photonics had not had the resources to put expert witnesses on the stand. But he planned to testify in the Control Laser case that was coming up in Florida. "A lot of false charges have been made that should be corrected," he said.

The laser industry rallied behind the idea that the real contest would be joined in Orlando, when the Control Laser case went to trial in September.

■■

The summer air in 1982 smelled heady and fragrant to Gould. His gas discharge amplifier patent was still pending, but his victory against General Photonics had led to his rediscovery by the media. Articles treated him as, if not *the,* at least one of the fathers of the laser. The journal *Science* had revisited the patent dispute in an April article, and Gould had held his own against the establishment.

Some saber-rattling by Bell Labs had been resolved to Gould's satisfaction. AT&T's research arm had filed a declaratory judgment suit—an action typically sought against a plaintiff by a potential defendant—asking that both of Gould's patents be declared invalid. Roy H. Wepner, who had been hired by the Westfield firm the year before, in part to help handle the Gould caseload, had countered with a suit charging Bell Labs with false advertising for claiming the laser had been invented there. The parties settled when AT&T, preferring not to walk into that nest of snakes, took a license on the patent at a concessionary rate of $150,000. Townes learned of the action in a phone call from a friend in Bell Labs' patent office. He knew it was a business decision, but he was disappointed anyway. Schawlow couldn't understand it, and would always say he felt betrayed.

A business reporter for *The Philadelphia Inquirer,* sent to report on Gould's status as the Control Laser case neared trial, found him living well.

Gould and Marilyn had moved again, this time to a small horse farm on five acres in Great Falls. The property had a swimming pool, stables, a riding ring, and it backed onto Great Falls Park, which had miles of riding trails in its lengthy stretch of Potomac River frontage. The house and land had cost $575,000. Gould, whose first memory was of riding on horseback up Troublesome Creek to the settlement school where his mother had taught deep in the Kentucky Appalachians, was in his sixties riding like a country gentleman.

"I am a wealthy man," he told the *Inquirer*'s reporter, Larry Reibstein, "and I guess I expect more (money) on the basis of the numbers. But if it all disappeared today, I wouldn't be unhappy."

The *Inquirer*'s article, which appeared in early July, depicted Gould as gaining ground. But Gould would have ample opportunity to test the equanimity he had expressed, for the summer idyll ended with amazing quickness.

40

Dick Samuel had taken Gould's gas discharge amplifier application to the Board of Patent Appeals. Failing there to have examiner Moskowitz's rejection of the application overturned, he had sued Patents Commissioner Gerald Mossinghoff in the U.S. District Court for the District of Columbia to compel the patent's issuance. He charged the Patent Office with treating Gould's application differently, not following regular Patent Office procedures but digging deep to find reasons to avoid granting Gould the patent.

On July 21, U.S. District Judge Thomas A. Flannery threw Gould's case out of court.

Flannery granted summary judgment to the Patent Office, meaning that he didn't need to hear evidence in a trial to reach a conclusion. He ruled in favor of the Patent Commissioner's argument that Gould's 1977 patent application had covered laser oscillators as well as amplifiers, after all. And because the interferences against Javan and Hellwarth had determined that Gould's original application, from which the 1977 application had derived, had not disclosed how to make an oscillator, he was prevented from pressing a separate amplifier claim. The ruling rested on a legal doctrine called collateral estoppal, meaning that once an issue is decided, and the decision upheld, the matter is over. Flannery's ruling closed, for the gas discharge amplifier, the door through which Gould had entered to win the optically pumped amplifier patent.

Samuel filed a motion asking Flannery to reconsider.

The Patent Office response told the Patlex lawyer that the PTO had made a decision to deny Gould's patents at all costs—not only those pending but the ones already granted. A motion filed by PTO solicitors said the Patent Office was trying to decide whether collateral estoppel applied to Gould's existing patents "to determine if those patents should be subject to reexamination." The motion practically begged for outside attacks on Gould.

In Orlando, Control Laser attorney Robert Duckworth immediately requested a postponement of the trial. He cited the Patent Office motion, and said Control Laser was going to ask the PTO to reexamine the optically pumped amplifier patent.

Samuel had rented an apartment in Orlando, and was living there while he prepared for trial. He had been there long enough that he'd established residency and could vote in the upcoming fall elections.

His trial preparations had been thorough. Looking at the case from all angles, Samuel had commissioned a juror attitude study aimed at learning what Orlando area residents thought about things like inventors, professors and graduate students, patent laws, royalties on patents, their role in manufacturing costs, and whether "much of the industrial progress of this country is based on the inventiveness of individual people." His Florida trial counsel, Warren Goodrich, had established the dress code that banned browns, vests, and loafers. Marilyn, before she and Gould left for Orlando for what they expected would be a lengthy stay, had demanded that Gould sit down with transcripts of his depositions from the interferences, and his testimony in the General Photonics case, to prep himself. Roy Wepner, who would share trial duties, went out and bought five new suits. Patlex's experts were waiting for word when they would be testifying. Charles Townes was champing at the bit to testify against them.

At the end of the week before the trial was to start on Monday, Judge George C. Young sent word that he wanted to see the attorneys in his chambers. Samuel arrived at the courthouse in downtown Orlando wondering what it was all about. Samuel suspected that he, and his beard, irritated Young. They were stylistic opposites—the intense, fast-talking New Jerseyan and the older, portly judge whose Southern accent managed to be slow and brusque at the same time. It certainly had taken Young no time at all to throw out Samuel's antitrust complaint against Control Laser just a few months after Samuel had filed it in 1978. When the attorneys all had gathered, Young dropped a bombshell into the trial preparations. He said he was recusing himself.

The judge cited a potential conflict of interest. Warren Goodrich was a partner at Holland and Knight, a prominent firm based on Florida's west coast.

Judge Young's wife was involved in a dispute that appeared to be leading toward a lawsuit, and he said Holland and Knight was likely to be involved.

Young's recusal, coming almost on the eve of the trial, tossed the case to District Judge Maurice Paul. The newly assigned judge looked at the complicated technical matters involved and postponed the trial indefinitely.

Samuel was furious. The Federal Express trucks that were on their way to Florida loaded with trial documents had to be turned around. Duckworth, for Control Laser, could not have been more pleased.

On the heels of the postponement, on September 9, 1982, the laser maker asked the Patent Office to reexamine the patent over which it was being sued. In a case of remarkable coincidence, Bell Laboratories filed for a reexamination of the optically pumped amplifier patent the same day.

Lumonics, facing Gould's infringement suit on his use patent, on September 29 asked the Patent Office to take a new look at that patent.

The Patent Office's virtual solicitation of reexamination requests on the Gould patents had found fertile ground. The PTO quickly agreed that "substantial new questions of patentability" had been raised by Control Laser and Bell Labs over the optically pumped amplifier. It granted both reexamination requests on November 16.

Three weeks later, on December 8, the Patent Office granted the Lumonics' reexamination request. By then, General Motors had filed its own request that the use patent be reexamined.

The reexaminations all found their way to the desk of a seasoned examiner named Harvey Behrend. Behrend was reputed to be one of the Patent Office's toughest. He went over applications with the tenacity of a terrier rooting out a mole, searching for gaps in their disclosure. He probed the applicable literature for any prior art an inventor might have failed to cite.

Assigning Behrend to the Gould reexaminations was a departure for the Patent Office. In the PTO's brief history in applying the new law, under guidelines set out in the Manual of Patent Examining Procedure patents usually had been returned to the original examiner, or to an examiner in the same unit. Behrend, however, had had nothing to do with either of Gould's patents. They had emerged from two different units, and he was in a third. His reputed toughness, and this change from normal PTO procedure, fed Dick Samuel's suspicion that the Patent Office was bent not only on denying Gould's pending patents, but snatching back the ones he already had.

Samuel took aim at the reexamination law, challenging its constitutionality in federal court in Pennsylvania.

Meanwhile, all action in the Control Laser litigation halted to await the outcome of the reexaminations. At the end of 1982, the summer euphoria was long departed and Gould's forces prepared for a new and costly round in the long war.

■

Gary Erlbaum wondered what he was doing at the helm of Patlex. Three years into the company's existence, he felt like a fish out of water. He was used to trying to make money, but with Gould's patents tied up he was an entrepreneur with nothing to work with. Patlex was little more than a client of the law firm. Helping to map a strategy and approach to patent litigation was not Erlbaum's long suit, and the business of keeping the company afloat while the litigation continued was not something that especially thrilled him. He could think of only one way to improve the company's efficiency, and he waited until the February 1983 board meeting to share it with his directors.

The board held its February meeting at the New York offices of the Chicago Options Exchange, whose chairman, Wally Auch, was one of the Patlex directors. The conference room meeting site offered a sweeping view of New York Harbor and the Statue of Liberty.

As Erlbaum walked around the conference table handing Gould, Langone, Samuel, and the other directors sealed envelopes, he talked frankly about his concerns. "The company is chewing up money in legal fees," he said. "We're being held in abeyance in every jurisdiction that we're in. If you'll open your envelopes, you'll find the first step to a solution. And then I'll tell you what I see as the second step."

The directors opened their envelopes and stared at Erlbaum's resignation.

"What I propose," Erlbaum continued, "is that you make Dick Samuel the CEO."

Samuel's shock was as profound as anybody's in the room. He was hearing Erlbaum's proposal for the first time, as were the rest of the directors. He started to object, but Erlbaum interrupted him.

"Hear me out," Erlbaum said. "The company's all about this litigation. If a litigator is in charge, it's going to save substantial money just in terms of strategic planning. I'm ready to move on, and I think it's the right thing to do for the company."

The other directors favored the idea, but Samuel wasn't sure. He went home and assessed his options. Samuel believed in Gould, but that didn't alter the financial reality. Lerner, David, Littenberg and Samuel was a successful, money-making law firm. Patlex was just one of many clients. Samuel would take a 20 percent pay cut if he resigned his partnership.

But Samuel had a big stake in Gould's case. It spanned much of his family history. His investment was not only financial, but emotional. His daughter by his first marriage was a twenty-one-year-old senior at Georgetown University. She had been twelve when Samuel had taken up Gould's patent war. He had told her about the forgotten inventor of the laser, and said, "Daddy took his case, and we're going to make a lot of money." Now he was divorced from his first wife, engaged again, and although the end seemed more distant than ever, he was not ready to let go.

Six years, he calculated. Even with the pay cut, he could continue for six years without going broke. He had come too far and believed too strongly in Gould's case to abandon the course now.

The next day, he informed the directors of Patlex that he would take the job.

■■

General Motors put both feet into the fray a month later, on March 4, 1983. GM was still waiting for word from the Patent Office on its November request to reexamine Gould's use patent. Now, it filed to have the optically pumped amplifier patent reexamined, too. The new request cited different new questions of patentability from the six already accepted by the Patent Office in the Bell Labs and Control Laser requests.

The weight of the forces against Gould could be counted in documents. Along with a 220-page "memorandum" and a check for the $1,500 reexamination fee—a laughable amount, considering the costs foreshadowed by the GM filing—the auto maker submitted a filing cabinet's worth of papers. They included five books of exhibits of the patents it claimed preceded Gould's, eight bound volumes of Gould's patents and his applications, five bound volumes tracking Gould's interferences through the Patent Office and the courts, two bound volumes of TRG contract documents, two bound volumes containing the transcript of Gould's depositions in his case against Lumonics and GM, five bound volumes of the exhibits accompanying Gould's deposition, a bound volume containing the transcript of the General Photonics trial in San Francisco, and more volumes and papers including depositions of Gould's witnesses in several cases.

It took six pages just to list the documents GM was submitting. Then there was a five-page table of contents to guide the reader through the ten-part memorandum. Already it was exhausting, without even getting to the documents.

Then, once you looked through it, there were the most incredible assertions of prior art. Gould had invented nothing new, GM contended, because light amplification had been detected by astronomers in the atmosphere of Mars.

GM's assertion that its request was in the public interest lay in the ever-growing use of lasers. Commercial optically pumped lasers now were a $100 million market annually. Lasers had become not just useful, but indispensable, in medicine, communications, and industry. Their defense applications were vital, and they had far outstripped Townes and Schawlow's first assessment of them as "interesting science," to become tools that underlay a vast range of other scientific work. Lasers had created a technological revolution.

But it was a precarious revolution, according to GM. The vast benefits conveyed by this unique machine might disappear if Gould's patent was allowed to stand. He could single-handedly kill off the progress created by his own invention. Manufacturers, optical surgeons, researchers, and communications companies might choose to go back to old ways of doing things, if GM's lawyers were to be believed. "The '845 patent threatens and imperils the continued unfettered use of optically pumped lasers in the foregoing publicly beneficial applications," they said, in a riot of legal hyperbole.

■■

GM wanted it both ways. While its lawyers were fulminating against Gould, the company had taken licenses on both the use and amplifier patents, and thus gotten itself dismissed as a defendant in Gould's patent infringement suit in Chicago. By taking the licenses, GM had limited its liability. It sought a further limitation by asking for the reexaminations; its royalties due under the use patent would be reduced by half if the amplifier patent was found in reexamination to be invalid.

The new law allowing reexaminations was supposed to provide individual inventors and small companies an inexpensive process for the defense of their patents. But that premise assumed a neutral reexamination by the Patent Office. The same law could also be used against them, as General Motors, Bell Labs, and Control Laser, and the Patent Office, were demonstrating. The weight falling on Gould was enough to crush most small inventors.

41

The reexamination requests all rehashed old history, which under the law they were not supposed to do.

Samuel H. Dworetsky, one of Bell Labs' stable of lawyers, revisited the preprint of the paper Charles Townes and Arthur Schawlow had written for *Physical Review* in 1958. He suggested that Gould had seen it before the date of conception awarded him in the General Photonics case, and therefore did not deserve the patent. He went back over the decision from the Townes and Schawlow interference that Gould's diligence did not begin until December 1, 1958, rather than the August date awarded him in the General Photonics trial, and made much of the uncontested nature of the trial.

General Motors' request on the optically pumped amplifier alleged that Gould had failed to mention articles by Basov and Prokhorov—published only in Russian—and Pringsheim's treatise on fluorescence and phosphorescence, among others, as source materials.

GM's request on the use patent was exponentially stranger. Here, the lawyers trotted out childhood science texts and scouting guides as "prior art." They said that because Archimedes, 2,300 years earlier on the island of Sicily, had suggested that sunlight could be reflected onto the sails of enemy ships to set them on fire, and because scout manuals taught boys and girls how to start fires focusing sunlight through a magnifying glass, Gould's suggestions for the way laser beams could be employed was nothing new, and he did not deserve the patent.

Alleging these documents as prior art was spurious on its face. But Dick

Samuel was fatalistic. He was convinced that the reexaminations would not treat Gould fairly regardless of whether the prior art submitted had any merit.

On tax day—April 15, 1983—Samuel's pessimism was rewarded. News of Behrend's first official action on the optically pumped amplifier patent filtered out of the Patent Office. The examiner, reacting primarily to Bell Labs' allegation that Gould had seen the Schawlow and Townes *Physical Review* article earlier than he'd previously claimed, rejected his most important claims to the optically pumped amplifer.

Patlex hurried to do damage control with the media and licensees. Samuel stressed that the patent remained valid while he went back to Behrend with new information. But if Behrend couldn't be convinced, he said, the company would appeal to the Patent Board of Appeals, and to the federal courts, if necessary. Patlex would go as far as it had to go to reverse the decision.

Gould pointed out, in an interview with a laser trade magazine, what by now was obvious. Behrend's decision perverted the intent of Congress. Rather than relieving him of the costs of litigation, the law was throwing Gould back into a lengthy appeals process that was likely to end up in the courts. Not only had the law not protected him from harassment, it had palpably increased it.

Gould didn't believe in conspiracy theories. He had fought on as long as he had largely because it never occurred to him that each new step in the arduous process would not be the last. It was hard for him, though, not to detect an ominous pattern in the Patent Office actions.

By now Behrend had granted General Motors' request, as well as Lumonics' earlier one, to look again at Gould's use patent. A week after his first rejection—a refusal to certify patentability—of the optically pumped amplifier, the PTO merged the use reexaminations into a single proceeding.

On May 13, it granted GM's request to reexamine the optically pumped amplifier patent, and would eventually merge that reexamination with the ones initiated by Bell Labs and Control Laser.

■■

The war now was being waged all out on two fronts. The reexaminations were only part of it. Judge Flannery in the U.S. District Court for the District of Columbia had refused to reconsider his decision supporting the Patent Office's refusal to issue Gould a patent on the gas discharge amplifier. Gould's legal team had taken the next step, appealing to the Court of Appeals for the D.C. Circuit.

William L. Mentlik was the point man on the appeal. He was thirty-four years old and energetic, with brown wavy hair and a preference for somewhat overstated gold jewelry. He had worked as a patent examiner in chemical arts while he attended law school at George Washington University, stayed on at the Patent Office, and later moved to the Justice Department's Patent Division before joining the ranks of private patent lawyers. Mentlik had the resilience required of anybody representing Gould, as he proved when he composed the appeal brief in his head, dictated it into a reel-to-reel tape recorder, and then erased it by mistake. The second time, he managed to deliver the tape to his secretary with the brief intact.

The appeal pointed out that collateral estoppel, used to deny Gould a patent on the gas discharge amplifier since the Javan interference had gone against him, applied only if the issue decided was the same. Therefore, the argument on which Flannery had decided for the Patent Office and against Gould was flawed. The case against Gould in the interference with Javan had decided only oscillators, while amplifiers were the subject of Gould's patent application.

A three-judge panel heard arguments on March 23. Mentlik had the sense that he'd done well, that after running into a series of stone walls Gould's case had finally reached some people who were listening. One of the judges questioned Mentlik's opponent, assistant PTO solicitor Fred McKelvey, in close detail. McKelvey, who had told Dick Samuel that if Gould ever won a patent on the gas discharge amplifier it would be the most valid patent ever issued, even complimented Mentlik afterward.

The panel's decision, when it came down in July, was a ray of hope in the year-long torrent of bad news. Different inventions were involved after all, the judges ruled. A case deciding oscillators did not apply to amplifiers, and therefore collateral estoppel did not apply. They reversed Flannery, and remanded the case to him for trial.

■■

By the fall of 1983, with Samuel at the reins of Patlex, Roy Wepner had taken over the direction of the legal work on the reexaminations. Wepner, who had a bachelor's and master's in mechanical engineering from Rensselaer Polytechnic to go with his law degree from Penn, was thirty-six and already getting gray after two years on the Gould cases.

Wepner had done his best to work with Behrend, trying one argument after another to persuade the examiner that the newly cited prior art had already been considered by the examiners who issued Gould's patents. Behrend wouldn't budge. Wepner became convinced that Behrend was a hit man assigned by the Patent Office to take out Gould's patents. He was intransigent, and Wepner filed a petition to remove him from the reexaminations.

The petition met a familiar fate.

"Surprise!" Wepner wrote Samuel sarcastically on September 20, 1983. "Our petition to disqualify Examiner Behrend has been denied."

Stuck with Behrend, Wepner had no choice but to try to deal with him. In the first stages of the reexamination process, the examiner was also a de facto referee over the various filings. Wepner filed an official "communication with the examiner," pointing out that the reexamination requests appeared to reflect a pattern of collusion.

Wepner was surprised, he said, that "to date, no government agency has looked into the 'public interest' aspects of the apparent collaboration between AT&T and General Motors, who are apparently working together to deprive Gould of the patent rights for which he has struggled almost continuously for nearly a quarter of a century." He suggested a close look at a series of actions:

"—General Motors picks a fight with Gould over the '436 (use) patent by intervening in *Gould v. Lumonics,* solely because Gould has sued one of General Motors' thousands of vendors.

"—AT&T picks a fight with Gould over the '436 and '845 (optically pumped amplifier) patent by filing a declaratory judgment suit, solely because Bell Labs is served with a deposition subpoena.

"—General Motors settles with Gould by taking a license under the '436 patent and receiving immunity from suit under the '845 patent, thus placing a ceiling on its potential liability.

"—AT&T settles with Gould under terms substantially similar to the General Motors settlement.

"—As trial of the *Control Laser* suit becomes imminent, AT&T [through Bell Labs] files a non-statutory request to re-examine the '845 patent.

"—As the possibility of a trial of the *Lumonics* case becomes real, by the sheerest of 'coincidences,' since AT&T has done its part by tying up the '845 patent, General Motors files for reexamination of the '436 patent.

"—As the reexamination of the '845 patent progresses, and the weakness of AT&T's arguments is exposed, General Motors steps in and files another request to reexamine the '845 patent.

"—Now, as the weakness of General Motors' position on the '436 patent is exposed, AT&T jumps in through Mr. Dworetsky's present letter."

The letter to which Wepner referred was part of the pattern—an apparent end run by AT&T to involve itself in the reexamination of the use patent. Procedures dictated that once General Motors and Lumonics had raised their original questions, they were barred from filing further arguments and had nothing further to do with the case. AT&T had not requested reexamination of the use patent. But it had sent to Gould, through his attorneys, an allegation that Gould might have seen the preprint of the Schawlow and Townes *Physical Review* paper earlier than he'd admitted. Dworetsky threatened to hit Gould with a fraud charge if he didn't file the allegation with the Patent Office as part of the reexamination record.

Gould didn't have to do that. Wepner, in his "communication" with Behrend, called AT&T's arguments "so pathetically specious that they would not be worth addressing even if Gould had any obligation to do so."

The Patlex lawyer had charges of his own to make, in any case. He had been trying for months, unsuccessfully, to pry from Bell Labs its preprint distribution list. He knew it contained only the names of Bell Labs' employees, and would deflate the charge that Gould had plagiarized material from the preprint. Bell Labs was still stonewalling, not only refusing to provide the list but not even acknowledging Wepner's requests for it. This, Wepner charged, amounted to "conscious and deliberate fraud."

He was in something of a state by then, and took aim at the "gross unfairness" of the reexamination process. "While Gould has already had to deal with truckloads of material and arguments filed by General Motors and Lumonics, and even though the regulations expressly state that the participation of requesters ends prior to the first official action, Gould must now contend with nonsensical arguments from yet another massive adversary. Clearly, this entire proceeding is a mockery of justice and should be vacated at the earliest possible time."

But any hope that the reexamination process could be short-circuited faded that November. Judge Edward N. Cahn in the eastern district of Pennsylvania ruled against Gould's attack on the reexamination statute. He said no issue was before the court because no one—so far—had been harmed.

██

On January 30, 1984, Wepner filed a request to interview Behrend. By then, he had responded to the examiner's first official action, and he wanted to know whether his answers had had any effect. The request was denied. Two months later, the Patent-Office granted the request.

Wepner, Samuel, and Gould went to the PTO offices outside of Washington on the afternoon of April 19 to sit down with Behrend and a supervisory examiner named Richard Schafer. They had a little more than an hour.

Samuel opened and then Gould explained how his disclosure differed from what Schawlow and Townes had written, while Wepner took notes. Gould genuinely wanted to be helpful. As far as he could tell, the examiner didn't really understand the difference between microwave and light frequencies, and what was wrong with one element pumping itself—you were left with only a weak line for pumping—so that it was necessary to use different elements for the pumping lamp and the lasing medium and find spectral lines that coincided. Behrend barely spoke at all. Gould could hardly believe it. The man just sat there. At one point Behrend asked Gould if the properties of fluorescent materials were known prior to Gould's invention. Well, of course, Gould said, there was Pringsheim. Then Behrend wanted to know how Gould had found out about the preprint, and Gould said he'd received it in printed galley form from one of the other physicists at TRG. That was it. Afterward, Gould couldn't remember how the examiner's voice sounded or even what he looked like.

He did remember thinking he could see Behrend's mind at work, churning away at how he was going to resist Gould's efforts. That was the point at which, in retrospect, he'd decided that nothing he said would make a difference.

Later, Wepner's detailed summary of the meeting consumed eight legal pages. Behrend's was three paragraphs.

Gould was right. The interview had no effect whatever. On July 6, Behrend issued his final official actions on both Gould patents. In each case, he rejected every argument Gould made, and maintained each of his grounds for rejection. It was astounding. Whatever you said to Behrend about Gould's patents, he just said no.

It was true, if you were entertained by this kind of thing, that he managed to say it in a variety of ways. And he used the phrase "Gould's arguments are unpersuasive" over and over. He disregarded the testimony of Gould's expert witnesses as unpersuasive, too. He wisecracked gratuitously here and there. He turned down another interview request, saying, "It is not apparent that an interview would serve to develop and clarify specific issues leading to a material understanding between the Examiner and the patentee and thereby advance the prosecution of the reexamination." He used Judge Flannery's opinion against Gould in the gas discharge laser case to support his rejections, even though the Court of Appeals had reversed Flannery and remanded the case for trial. He bought the argument that General Photonics hadn't put up a defense, and so gave no weight to Gould's victory there. Nor did he give any credence to the work the previous examiners had done in issuing Gould's patents in the first place. Behrend simply adopted each and every issue that

Control Laser, Bell Labs, and General Motors had used to attack the amplifier patent, and each one that Lumonics and General Motors offered against the use patent, and in some cases he even incorporated their dismissive and sarcastic language by reference into his opinions.

Gould and Patlex appealed the optically pumped amplifier case to the PTO's Board of Appeals. That, clearly, was where the Patent Office wanted it to go. Meanwhile, Wepner continued to press Gould's arguments with Behrend over the use patent, but with little hope of changing the result.

■■

Just when it appeared the Patent Office couldn't do any more damage, Gould got notice that his Brewster's angle window application had been reassigned to Behrend.

Wepner protested the reassignment. He thought it was abundantly clear that Behrend was incompetent in his grasp of the technology, and was biased against Gould to boot. In its response, the PTO inadvertently revealed the shocking lengths to which it had perverted the reexamination law to take away Gould's patents. The Patent Office cited "conservation of Office resources, consistency in patentability determination, and avoidance of conflict in Office policy determination" in giving the application to Behrend. It went on to say that Gould "has received, and will continue to receive, an objective, fair, and impartial examination of each of his pending patents."

But Gould had only one patent pending, the Brewster's angle application. The gas discharge amplifier was out of the PTO's hands in the courts. The other two patents—the optically pumped amplifier patent and the use patent—existed. They were real, and would remain so until proven otherwise. At least that was the way it was supposed to work. Maybe the wording was just a weird Freudian slip, but it seemed to Gould and the lawyers that the Patent Office was treating the patents in reexamination not as existing patents that already had been thoroughly reviewed, but as new applications that had to be justified all over again.

42

Toward the end of 1984, Ken Langone revisited an earlier demand that Gene Lang and Refac turn over more of the royalty pie. Either that, he told the licensing company's chairman, or have a change of heart about paying for the patent litigation.

The situation rankled Langone. It had been gnawing at him for some time now, the fact that Refac had 40 percent and was bearing none of the extensive legal costs that kept going up with every brief, filing, memo, communication, and appeal. The more Langone looked at it, the more it looked like a free pass. The blunt-spoken investment banker saw Gould's patents mired in a swamp of litigation, in which Patlex was doing all the work and spending all the money. If Gould won, Patlex would get its 40 percent. But another 40 percent of the cherries would fall into Lang's basket, and he wasn't doing much to shake the tree.

Plus Langone was preparing Patlex for another public offering, and he didn't think Patlex's share would look good enough to lure investors.

Lang begged to differ, naturally. As far as he was concerned, he'd been working hard to sell licenses, beating his brains out, in fact, to find laser makers who were willing to buy licenses on favorable terms in anticipation that Gould might actually win. Because if they waited, the terms would be much different. And he had been mildly successful, producing a small stream of royalty income that so far had totaled about $750,000.

It still bothered Lang that he'd given up a portion of his original share, and he was quick to rebuff Langone's suggestion that he give up still more.

"We're putting the money in," Langone argued. "If we're going to continue, we have to have more shares. If you want to sit there and not participate, then you've got nothing as far as I'm concerned. We're putting the only real money in the game. Unless we get an additional economic interest, we're not going to keep doing it."

Lang still said no.

The relationship between the partners had been deteriorating for some time over the way the shares were split and the cost of litigation. The bad blood went back at least five years. That was when Lang had arranged the *Business Week* interview with Dick Samuel and Joe Littenberg at his office, the one in which Samuel bragged that they had invented the laser patent application. Lang thought he had done the enterprise a favor, and so had been astounded when the lawyers billed Refac for their time.

But there was no getting over the ongoing expense of the several litigations. Langone pressed his case ever more forcefully. Finally, on December 4, 1984, Lang signed a deal in which Patlex acquired all of Refac's remaining interest in Gould's patents in exchange for Patlex stock and an ongoing 16 percent interest in the patent royalties.

If Gould's patents ever became the cash cow his investors hoped, Patlex would get 64 percent of their revenue, Gould 20 percent, and Refac 16 percent.

At the same time, Patlex acquired from Refac the ongoing right to negotiate licenses on the Gould patents. Samuel was behind that part of it, and there was in the move a message for Gould's enemies if they had chosen to take notice. It was that Patlex was impatient, and was going to play hardball with anybody who persisted in ignoring the Gould patents. The deal amounted to a 16 percent payoff to get Refac out of the picture and free Patlex of a partner it didn't want and felt hampered by.

■■

On the morning of Monday, January 28, 1985, Gould and Marilyn got up early and drove from suburban Great Falls to downtown Washington to meet Dick Samuel, along with Roy Wepner and Sid David, one of the law firm's senior partners, at the L'Enfant Plaza Hotel. From there they took a cab to the federal district court building on Sixth Street off Constitution Avenue, across from the National Gallery of Art. It was a bright, cold day, and the trial that the Court of Appeals for the D.C. Circuit had ordered Judge Flannery to hold in the gas discharge amplifier case was about to begin.

The courtroom was large, modern, and impersonal. It seemed to swallow the participants in the non-jury trial. Samuel and the other lawyers had been in Washington since the week before, preparing and discussing the issues they wanted to pursue. Would they lean on evidence, or legal points? They knew the Patent Office intended to use TRG's progress reports to the Air Force on Project DEFENDER to argue that Gould had not made a working laser.

Flannery looked vaguely Irish, not in the clichéd rosy-cheeked, hard-drinking way but substantial and middle-of-the-road, so that if you found yourself beside him on an airplane flight you wouldn't be surprised to learn he was a judge. He made it clear that even though he'd been reversed, he remained inclined toward the Patent Office's position. "I still think you guys have got a problem here," he told Gould's side.

But as the trial went forward, that seemed less and less to be the case.

Patent Office solicitor McKelvey opened by arguing that Gould was "trying to change history." Marilyn, sitting in the spectator seats behind Gould and the lawyers, began to experience a serious burn. She had never heard a more blatant example of the truism that history is written by the winners, who didn't like to be corrected. "What history?" she thought. "History written about a laser patent that turned out not to work? History according to people who didn't want to believe what Gordon did and companies that didn't want to pay him? Yeah, we're trying to change history. We're trying to make it right."

Gould was the first plaintiff's witness. He was a veteran by now, having testified in the General Photonics trial and given countless depositions and been examined and cross-examined by the equivalent of a small city's entire Bar Association. Gould had explained himself over and over again, ad nauseam, and he had a pretty good handle on the way the game was played. The only thing that bothered him about it, to be honest, was that while he was on the witness stand he couldn't smoke. He was going to have to quit one of these days anyway. A few more months, and he'd be sixty-five.

Sid David took Gould through the process by which he'd come up with the gas discharge amplifier. Gould talked about his intense study of sensitized fluorescence and lasing possibilities when he moved from Columbia to TRG, and his realization that when there were transfers of energy between colliding atoms of two different elements—collisions of the second kind—and those energies corresponded to visible wavelengths, amplification was possible. He wrapped his raspy smoker's voice around the simplest terms possible, because Gould was also used to lay people not being—how would you put it?—quite comfortable with the physical details.

The PTO came out just as Gould's side had expected, relying on TRG's reports to bolster their case that Gould's ideas hadn't worked. They were the same documents used against Gould in the Hellwarth interference, where Hughes Research attorneys had made their case Gould couldn't have invented the Q-switch if the laser hadn't been invented yet.

What Fred McKelvey, the PTO solicitor, seemed to have forgotten was that the laser wasn't the issue anymore. It didn't matter whether TRG had struggled to build a laser oscillator using Gould's specifications. What lay at issue now was whether Gould's ideas for achieving amplification, the vital first step to laser oscillation, had been realized.

Gould called two other witnesses, Samuel, and Peter Franken, now director of the Optical Science Center at the University of Arizona. Franken was a persuasive expert witness as he testified to what persons skilled in the art would have made of Gould's patent disclosure when it was filed in 1959. They would certainly have been able to use it to build a laser amplifier, he said, and would have known which of Gould's specifications in the application referred to the gas discharge as opposed to the optically pumped device.

Franken was apt with an analogy as well. McKelvey latched onto a piece of the TRG report that referred to amplification without net gain.

"How can there be amplification without gain?" McKelvey asked, feigning incredulity.

"Take a money market fund," Franken replied, his face displaying the same elfin humor and his voice conveying the same energy Gould remembered from their graduate school days. "Say its value increases. That's gain. But there may also be losses, such as front end charges or back end charges. With those losses, you may end up without any gain. But that doesn't mean there hasn't been amplification."

The Patent Office continued to press its case that Gould hadn't taught how to build an amplifier in this original patent application. McKelvey called Gould as a witness, too. He wanted to ask about a sodium-mercury gas discharge laser Gould had paid Optelecom to build the year before, just to show his patent application was enabling. The idea was that Gould had to stay out of the lab to prove that the machine could have been built from his specifications alone. He had. McKelvey pressed to get a different answer, but Gould would not go where McKelvey wanted to take him, and the solicitor just gave up. Also on McKelvey's witness list were a forty-year-old physicist from the Naval Research Labs who was expected to counter Franken, and Gould's former boss at TRG, Dick Daly.

Daly's testimony was amusing. Overlooking the fact that he was not a disinterested witness, since his company, Quantronix, was among the firms Gould was suing for infringing on his optically pumped amplifier patent, Daly's memory seemed to be faulty. At issue was what TRG did in its gas discharge laser experiments. The scheme Gould described most fully in his patent application, sodium-mercury, was never tried, nor was helium-neon, the combination Javan and Bennett had demonstrated at Bell Labs. Daly was asked what happened to the third, krypton-mercury.

"It was promising," Daly said. "But it didn't work. If it had, we would have followed it up. We followed up everything that indicated possible success. We would have pushed a peanut a mile on our hands and knees to make it work. We would have been crazy not to."

But hadn't there been a gain? Sid David asked. He put a paper under Daly's nose and asked him what it said. Daly studied TRG's year-end report in 1960 and his face under his bald dome grew red as he read the conclusion that "inversions and amplification can be obtained." His memory jogged, he then remembered that TRG had put krypton-mercury aside not because it didn't work, but because it wasn't thought powerful enough. "It remains our intention to pursue this scheme further when funding and time pressures are relieved," said the report. "There is a value to the gas discharge scheme in its potentially narrow emission line."

That pretty much sealed up things. Nothing the Patent Office's expert, a nice man named Barry Feldman, said could undermine Franken's knowledge as a person skilled in the art in 1959, when Gould filed his application. Feldman had been fifteen in 1959, and had talked to laser pioneers such as Gould and Franken only in ensuing years.

The Patent Office had begun its case asserting ten reasons for denying Gould a patent. Five were dismissed before trial as a result of the Court of Appeals ruling. The Patent Office abandoned three more at the end of the trial. When Flannery sent everybody home to await his opinion, the PTO was claiming only two defenses—that Moskowitz, the examiner, and the Board of Appeals, had denied the patent for adequate reasons under patent law, and that Gould had filed an insufficient disclosure by not providing enough information to build a working amplifier.

Gould left at the end of the six-day trial feeling as if his side had made its case and made it well. Still, he couldn't be sure, given Flannery's opening comments. There was nothing to do but wait.

43

The months following the trial were unbelievably tense for Gould. Nothing was resolved. He was feeling the aches of age as never before, and tremors in the unshakeable optimism that had made him think that each new barrier would be the last. Looking back and taking stock, he found it almost beyond belief that he had come more than a quarter of a century from his original invention. He felt like the legendary Sisyphus, pushing his rock up the mountain and almost reaching the top, only to have it escape him and roll back down again.

It was complicated. Gould had had cataract surgery after the trial. The surgery on the first eye had not gone well at first—the acrylic replacement lens had turned sideways in his eye, and had itself to be replaced. When Gould was assured that the new lens had taken, he went ahead with the second eye. He was in recovery when he noticed a shadow beginning to fall across his vision, and he knew immediately that his retina was detaching from the back of the eye. "Hey," he called out. "Somebody get in here."

By some miracle of coincidence, a young laser eye surgeon named Frank Spelman shared the office with Gould's cataract surgeon. Within two hours, Gould was back in the operating room for a vision-saving reattachment of the retina. The coagulating power of the laser's heat tacked the retina down around the edges like so many tiny rivets. Gould wasn't awake for that operation, but he was later when Spelman was touching up his work. Gould held his eye open and stared into the pulsing argon laser's reddish light and thought to himself, "What if this machine didn't exist? What if no one had invented it?"

And that's why it was complicated. The waiting was more frustrating than it might have been otherwise. Gould could not only see the laser all around him, he could see because of it, and that forced him to remember that he still was hostage to the Patent Office and the courts.

And now, bad investments in the stock market had eaten deeply into the payments he'd received from Patlex. Looking forward to retirement, but unsure if he would ever be able to say goodbye forever to financial worries, he had started paring down.

By the summer of 1985, Gould had sold most of his 25 percent stake in Optelecom, parceling it out in dribs and drabs over the previous year. The stock sales had paid him about $500,000, but with his losses on market speculation and the high interest he was paying on the mortgage on the Great Falls horse farm, he was far from secure.

When Marilyn stopped working that summer, and Gould told Bill Culver he was ready to retire from his daily laboratory supervision at Optelecom—though he would remain on the board—they decided they no longer needed to live so close to Washington. They put Matildaville Farm, as they had named the little horse farm, up for sale and scouted new locations. They wanted to be near the water again, and looked on Long Island's North Shore and along the Connecticut coast. Somehow, though, those areas had changed during the time they had been in Washington. The real estate boom of the 1980s had sent prices soaring, and living on Long Island demanded an outsized premium. Connecticut had gotten crowded to the point of claustrophobia. Meanwhile, the farm sold, and Gould and Marilyn were forced to take a rental in Great Falls.

Marilyn continued to browse the ads, however, and no sooner had they signed the lease than her eye fell on an advertisement for a house near Kinsale, Virginia, on the thrust of land between the Potomac and Rappahannock rivers Virginians call the Northern Neck.

They drove from Washington to take a look. The one-hundred-mile drive took two and a half hours, during which the traffic fell away and the buildings got lower and more sparse and houses retreated from the highway to sit discreetly back, and then disappeared from view altogether. The Kinsale house lay on twenty-three acres. It took some driving to reach the property after they left the main road, and another half mile bumping along a dirt track to reach the house. It proved to be a modest gray ranch, but comfortable, with a large porch across the back where they would be cool on the hottest days and dry in the middle of a rainstorm.

The property had 1,800 feet of waterfront on a Potomac estuary called Bonum Creek. The Potomac was wide there; the opposite bank was nearly out of sight and just a little way downstream the river emptied into the Chesapeake Bay. There were no neighbors within half a mile. At $285,000, the place was affordable, and it came with a boat, a dock, and a crab pot in the creek.

Best of all, it was utterly serene. Gould and Marilyn debated whether they wanted to be that far from the music and culture that they treasured. Plus Gould would be driving to Washington for Optelecom's board meetings, and they both would be flying to New York and Philadelphia for meetings with Patlex and the lawyers. But they also needed a retreat from the patent wars, a place to recover and spend some time at peace. And so that fall they moved to Bonum Creek, where they adopted a routine like Kenneth Grahame's Water Rat, simply messing about in the sixteen-foot outboard runabout that came with the house. From the water, they explored the banks of the creek and the Potomac, dodging the thickets of crab pots set by the local watermen and watching ospreys nest in ancient trees. They took long, restorative walks in the open forest on the property, accompanied by Marilyn's oversized and sprawl-ingly affectionate dog Angie, a water-loving mixture of German shepherd and Labrador retriever, and the spavined, gimpy Gammon, a shepherd which in her old age had developed the sense to not go near the water.

■■

"Why is he taking so long?"

That was the question Marilyn asked Dick Samuel as the fall of 1985 deep-ened into winter with no word from Judge Flannery. They were talking after a Patlex board meeting in New York. In August Behrend had taken the next step toward actually voiding Gould's use patent, filing an eighty-two-page opinion that cited prior art for his "refusal to certify patentability." Marilyn, and Gould, needed some good news.

"It must be bad news if he's taking this much time," she added, applying the conventional wisdom of television courtroom dramas.

"You can't go by that," the lawyer said. "That's just the way judges do things. They take their time. And it could be good news, or bad, you never know."

Less than a week before Christmas, on a Thursday, the phone rang at the house on Bonum Creek. It was Gould's accountant, calling to report that Gould was getting a substantial tax refund. He and Marilyn were discussing this when another call came in. This time it was Bill Mentlik at the law offices in Westfield, New Jersey. He'd just received a call, he said, from Flannery's law clerk. He couldn't say too much, because he hadn't seen the opinion. It was forty-five pages long, and it was supposed to be faxed to him shortly. But the word was, it was favorable, very, very favorable.

Gould and Marilyn resisted celebrating. They refused to believe that the long stretch of bad news was finally over.

But not much later—in the time it took the facsimile machine to spit out its burden of pages—Mentlik called back. He sounded incredibly excited. "Listen to this. Listen to this," he said. "Let me just read you the last part." Gould and Marilyn, both with phone receivers to their ears, could hear shouts and exclamations in the background. "Judgment is hereby rendered in favor of plaintiff Gordon Gould and against the defendant, the Commissioner of Patents and Trademarks. The defendant is hereby directed forthwith to issue to the plaintiff a United States patent including claims one through fifteen as set forth in U.S. Patent Application No. 823,611 insofar as they relate to a gas discharge amplifier."

"I'll be damned," said Gould.

■■

Mentlik and Roy Wepner took turns on the phone reading from Flannery's opinion. It was a complete reversal of his previous embrace of the Patent Office's position, a U-turn if ever there was one. The judge had dismantled the Patent Office's defenses. He had written, "The court is thoroughly convinced that the PTO made several material errors in determining that Gould's disclosure was insufficient . . ."

With collateral estoppel out of the way, removed as an issue by the Court of Appeals, Flannery had easily seen that Gould's application described gas discharge amplifiers and that his description was enabling—somebody who knew what he was doing could have built one. His ruling was a road map to the long, sorry history of the Patent Office's handling of Gould's application after the outcry that followed his first patent. It had repeatedly, from the examiner through the Board of Appeals, misapplied oscillator rulings to withhold a patent on the amplifier. One error was Moskowitz's strangely telling holding that the amplifier was both insufficiently disclosed and obvious at the same time. He had gone on to draw the wrong conclusions from the ruling in the Hellwarth interference, asserting that it applied to amplifiers when it did not and even adding language to that effect where none existed.

Flannery put it bluntly. The examiner, he said, "had no evidentiary basis to question the adequacy of Gould's disclosure, and Gould's disclosure should have been accepted as presumptively enabling."

The board had erred similarly, misreading dates to infer that Gould's application required too much experimenting, failing tests of discernment about what laser physicists would have known when the first lasers were being built, and drawing the wrong conclusions from expert testimony.

The judge pointed out that Dick Daly had testified in 1968 that Gould's notebook would have been enough to build a gas discharge laser using collisions of the second kind. At the time he gave that opinion, Flannery continued, Daly had not been sued by Gould for infringement and there was no reason for his opinion to be other than impartial. He used that background to credit Daly's 1968 testimony "over any implication in his present testimony to the effect that Gould's patent application contains an inadequate disclosure."

On the contrary, Flannery wrote, Gould's application "contains a written description of the claimed invention and of the manner and process of making and using it, in such full, clear, concise, and exact terms as to enable any person skilled in the art to which it pertains, or with which it is most nearly connected, to make and use the same."

It was a slam dunk, but there was more. "The court accords more weight to the testimony of Dr. Franken than to the testimony of Dr. Feldman," the judge wrote. "Dr. Franken's credentials are more impressive. His manner of testifying and the reasons given for his opinions have convinced the court to accept his expert testimony over the conflicting expert testimony of Dr. Feldman."

Finally—and here was the point that Gould found most gratifying, for it tended to support the quality of his early thought about the laser—the judge had written, "The PTO has not demonstrated that any individual ever made a serious attempt to build a gas discharge laser amplifier with any combination proposed by Gould and failed." Not only had Gould filed an operative disclosure, he had filed no inoperative ones.

Dick Samuel was in Israel on a business trip when Flannery released his ruling. Wepner called him in Jerusalem as he was on his way out to dinner. Returning to his rental apartment at the end of the meal, he called New Jersey and insisted on a line-by-line reading. It went on for awhile. Samuel had to stretch out on the kitchen counter to save his aching back. The phone bill came to $700.

Marilyn Appel had absorbed much disastrous bad news during her years with Gould. She had formed the habit of following each reversal on the patent front with a celebration dinner, a way she and Gould could shake their fists at the storm together. The dinners weren't elaborate, a carefully roasted chicken, say, and some accompaniment, but always a spectacular white wine. That's what she did on this night of a real celebration, she and Gould caroming around the kitchen in their isolated home, scarcely containing their excitement and producing a meal that they washed down with a Batard-Montrachet they'd kept on hand for just such an occasion. Marilyn had had to wipe the bottle clean of dust when she put it in to cool. "Frankly, Gordon," she said to Gould later as she surveyed the rich golden liquid in her glass, "I was afraid we were going to have to drink this under much worse circumstances."

■■

The phone kept ringing the next day. The press had gotten wind of the decision, and wanted Gould to comment. Wall Street already was responding, with a surge that was pushing Patlex's share price from $5 to $10.25 at the end of the trading day and making it the NASDAQ's steepest gainer.

The financial reckoning rested on the fact that gas discharge lasers accounted for some 60 percent of the lasers manufactured in the United States. They had outstripped solid-state ruby and other optically pumped lasers and become ubiquitous in supermarkets as optical scanners relieving checkout clerks of keypunching duties. They were the lasers used in compact disc players to read the music discs that were displacing cassette tapes. Lasers were edging into every home, in addition to their well-established uses in science and industry, and in medicine including eye surgery. Flannery's ruling came at a time when gas discharge lasers were a $300 million a year market.

Gould, demonstrating a penchant for the obvious, told *The Washington Post* that the ruling was "an important victory," and observed that "the wheels of justice grind very slowly." He predicted that the ruling would speed the remaining litigation and reverse the psychology of "the big companies that have been trying to pretend that I didn't exist."

His most prescient comment, though, was that the ruling "doesn't mean that they'll automatically pay up."

■■

There was one flaw in Flannery's ruling from Gould's point of view, and it was significant. The judge had no power under law to order the Patent and Trademark Office to issue a patent. It was a thread, and the Patent Office grabbed it. The order was one of three grounds the PTO used to appeal the case back to the higher court. It also argued that Gould's evidence wasn't sufficient to overturn the examiner's rejection, and that Franken's testimony should have been thrown out because it was tainted with knowledge about lasers developed after 1959.

The Patent Office was demonstrating once again that it would go to extreme lengths to deny Gould the patent, and if it had to issue it, would do so only grudgingly.

44

Gould and his two neighbors on Bonum Creek leased much of their land to a farmer, who planted corn one spring, soybeans the next, and each year, winter wheat. From December to March, the wheat formed a luxuriant green carpet that Gould and Marilyn could see from their front windows. Then it rose with the temperatures to become a tawny beard almost two feet high before the late spring harvest.

The wheat was cut and the new crop planted when Judge Flannery responded to a motion that accompanied the Patent Office's appeal. The judge was impatient. The PTO wanted a stay pending appeal, but Flannery slapped the motion down in blunt language.

First, he said the government was not likely to succeed on the merits. Then he wrote, "While the government argues that it will be harmed if the stay is not granted since the appeal may be rendered moot once a patent is issued, the harm to plaintiff appears much greater on balance. It has been nearly thirty years since plaintiff, now age 66, applied for the patent involved in this case. Plaintiff also notes that two companies have attempted to infringe on the prospective patent idea and that issuance of the patent is essential to plaintiff's protection of his rights.

"Finally, the court notes that there is a strong public interest in the timely award of patents, a point which the government does not dispute."

Flannery's impatience signified a sea change, and it was not much longer coming.

The corn on Gould's farm plot was eight feet high and ready for the crib on September 26 when the PTO's Board of Appeals issued a stunning ruling.

Harvey Behrend had cited eight reasons for rejecting Gould's optically pumped amplifier patent on reexamination. Examiners-in-chief Saul I. Serota, Ian A. Calvert, and Paul J. Henon, comprising the board, ruled that Behrend had failed completely to make his case, and reversed him on each count.

Patlex by then had shifted its headquarters west. Since winning the General Photonics case and taking over the company when it couldn't pay back royalties, Patlex had expanded in the laser field. It had purchased a manufacturer called Apollo Laser in a stock deal, and now used its offices in the San Fernando Valley town of Chatsworth, California, as its base of operations. Dick Samuel was on a golf course in nearby Northridge when he spied a cart approaching at top speed over the rolling terrain. Soon he recognized his wife and his secretary, and he wondered what could have prompted such a breach of golfing etiquette. One of the women was waving what looked like a sheaf of papers. Then he noticed their expressions, and relaxed.

The news was not universally good, as Samuel and the lawyers in Westfield knew as soon as they had reviewed the information from the Patent Office. Two boards, composed of the same members, had released reexamination rulings simultaneously. While Gould had scored a clean knockout on the '845 optically pumped amplifier patent, the ruling on his '436 patent on uses of lasers fell short.

Here, Behrend had cited five grounds for refusing to certify patentability. The board reversed the examiner on four of them, but let one stand. Ironically, the single prop shoring up Behrend was Gould's own United Kingdom laser oscillator patent, issued in 1964. According to the examiner, the UK patent had anticipated uses of lasers in a way that Gould's original application had not, and therefore he was not entitled to his original filing date of 1959.

The board's affirmation sent Gould's lawyers back to the U.S. District Court, appealing on grounds that Behrend, and the board, exceeded the scope of the reexamination law by considering something other than prior art to reject the patent.

But the big news was the certification of the '845 optically pumped amplifier patent. The trial in Orlando had been stayed for four years while the reexamination process worked its slow way to conclusion over the patent at issue in the suit. Now the final impediment to the trial had been removed, and Control Laser was out of excuses.

The case now lay in the hands of U.S. District Judge Patricia C. Fawsett, a former civil litigator whom President Reagan had appointed to the bench just a few months earlier, in June. She had a crowded trial docket, and the first opening for what promised to be a lengthy trial remained some time away. The parties looked to the fall of 1987 to begin what promised to be the deciding battle in the patent war.

Other skirmishes commanded Gould's attention in the meantime.

▐▐

During his days at TRG, Gould had worked with Ali Javan's gas discharge collaborator William Bennett and TRG colleague William Walter to develop a laser using copper vapor as a working medium. They had filed a patent application on the laser in 1965, and the patent was issued in 1971. The copper vapor laser was one of a class of metal vapor lasers that could operate only on a pulsed basis. It produced a spear of light that was bright green, right in the middle of the spectrum. Its drawback was the cost of building equipment to heat copper to a vapor at 1,400 degrees centigrade, but the laser's uses, while limited, made it worthwhile. A San Francisco Bay area company, Cooper Lasersonics, and a subsidiary, Plasma Kinetics, manufactured them for high-speed photography for scientific purposes. The patent was one of those Gould had bought back from TRG's successor, Control Data. It had less than two years to run when Gould and Patlex sued the two companies for infringement.

The copper vapor laser case moved to trial quickly, with none of the delays that affected Gould's broader and more controversial patents. Because it was the first of Gould's patent cases that would reach a jury of lay people, Gould's lawyers treated it as a runup to the pending Control Laser trial.

Dick Samuel prepared the case, but it was tried by Roger L. Cook of the San Francisco patent firm Townsend and Townsend, which Patlex had hired as its local lawyers in keeping with federal procedure. The relationship went back to the General Photonics trial, where Cook had negotiated the infringement settlement. U.S. District Judge D. Lowell Jensen presided.

It was Warren Goodrich's idea, backed by Samuel, to employ a phantom jury for the trial. These were local people hired to sit in the courtroom, follow the testimony, and then report their impressions. The lawyers wanted to make sure jurors grasped the daunting technology involved as well as the Byzantine twists of patent law. They also wanted a day-to-day picture of how they were doing—audience feedback about their presentations, courtroom manner, and the clarity and direction of their arguments. Gould's side brought together a group of candidates and employed four for the duration of the trial.

Gould, Marilyn, Samuel, and a contingent from Patlex virtually took over a small hotel, the Orchard, near Nob Hill. It was a little overdone, with ornate gilt mirrors and crystal chandeliers and great big mahogany armoires and chairs that looked like they'd been plucked from a Victorian estate sale, but it gave them what they needed, which was a base of operations. They repaired there after each day's proceedings. Marilyn took the phantom jurors to an upstairs suite to interview them, and then reported the results to Cook and Samuel. That shaped the next day's strategy, allowing the legal team to correct its missteps and shift its emphasis.

The phantom jurors found Gould charming and avuncular during testimony that covered several days. When it came time for the other side to present its case—the defense carries the burden in an infringement trial to prove the patent-in-suit is invalid, or that it's not infringing—Cooper Lasersonics called a witness whose name caused Gould to do a double take. It was Irwin Wieder.

Gould hadn't thought much about Wieder in recent years, since Gould had prevailed in the interference over the optically pumped amplifier. One of more than twenty claims in Wieder's patent had survived, but it had no commercial applications and Wieder had faded to secondary status on the laser scene since Westinghouse's bullish—and short-lived—announcement of his patent. Now Cooper Lasersonics had him testifying as an expert witness.

You would have thought Wieder had waited for this opportunity for years. Cooper Lasersonic's lawyer, Tom Herbert, a tall, imposing man with a voice that was big in the courtroom and soft out of it, took Wieder through the paces and Wieder said Gould had never invented anything. Meanwhile, Roger Cook was salivating.

When it was Cook's turn, he did a better job of pumping up Wieder's resumé than Herbert had. He asked the physicist about his award of a significant patent, about the coverage in *The New York Times* and other national papers, about the awards he had received from Westinghouse in gratitude. He had Wieder positively glowing. Then he said, "But wasn't there an interference?"

Wieder said yes, there had been an interference, and Cook asked him to explain to the jury what an interference was. Wasn't it sort of like a little trial inside the Patent Office? Wieder agreed, and Cook wondered what was the result. Wieder grumbled that except for a single claim, he had lost it on priority to another inventor.

"And who was that inventor?" Cook queried. By now he was having trouble getting Wieder to speak up.

"Gordon Gould," Wieder mumbled.

"I'm sorry," Cook said. "I didn't quite catch that."

Wieder repeated Gould's name a decibel louder.

"I'm sorry, but I don't think the jury heard you. What was that name again?"

"Gordon Gould," Wieder bellowed, rising from his chair with his face reddening, arm extended and a finger aimed in Gould's direction. "*The* Gordon Gould, that man right there."

The trial took a month, during which Gould's team learned from the phantom jurors, among other things, that Cook had jingled the change in his pocket during his final argument. When it was over, the six jurors ruled for Gould and said that Plasma Kinetics and Cooper Lasersonics had infringed his patent willfully—a decision that opened the door to triple damages. The judge overruled the jury on willful infringement, but barred the laser makers permanently from infringing, and let the royalty rate stay at the 6 percent set by the jury. The award to Gould and Patlex totaled $160,000.

The companies paid the judgment, then closed their doors for the time it took Gould's patent to expire.

Also in May, the Canadian laser maker Lumonics settled out of court with Patlex and resolved its infringement suits.

■

Close on the heels of the San Francisco victory came another, more significant, one. The Court of Appeals for the Federal Circuit on June 25 turned aside the Patent Office's last-ditch effort to deny Gould a patent on the gas discharge amplifier.

Circuit Judge Jean Galloway Bissell, writing for a panel that included Senior Judge Marion Tinsley Bennett and Judge Glenn L. Archer, Jr., described Gould's tribulations in her sketch of the background to the case. "The application in suit arrives at this court after a long, arduous journey through the patent continuation, division, and interferences practices in the U.S. Patent and Trademark Office, starting with an application filed on April 6, 1959." The description only hinted at the vicissitudes of Gould's long journey.

The Patent Office had cited three grounds in its appeal, and the appeals panel upheld Flannery on two of them. It said Gould had indeed presented evidence to overcome the examiner's case for rejecting the patent, and that Peter Franken had known and testified to the state of the art in lasers in 1959, without resorting to subsequent knowledge.

Where Flannery had erred, said the panel, was in ordering rather than authorizing the Patent Office to issue Gould a patent. It sent the case back to the District Court and told Flannery to amend his order. The remand added another step to Gould's ordeal, but the Patent Office was out of appeals. It had no excuse now not to finally issue Gould a patent for gas discharge amplifiers.

45

Gammon, the older, lame dog, had spent most of her life in a stable dodging kicking horses, not always successfully. She had been an unpaid and underappreciated stablehand, and she was ready for an improvement by the time Gould and Marilyn took her. Life at home was all she could have asked, but she never minded entering a kennel, either. It was like a spa vacation—plenty to eat and drink, limited exercise, and people waiting on you hand and foot.

So she was not at all unhappy when Marilyn dropped her off a few days after Labor Day 1987. She sidled happily into her big cage, bent tail wagging, and gave her particular dog grin to show it was all right when Marilyn said she'd be back in a month or so. An hour or so later, Angie's size was challenging Delta Air Lines' cargo handling capabilities, and the family was off to Orlando for the Control Laser trial.

Angie, coal-black and on the exuberant side, in addition to weighing a hundred and fifteen pounds, was not what the reservations clerk had had in mind when she had told Marilyn the Radisson hotel near Ivanhoe Lake took small dogs. But nobody said anything when they arrived. Between Gould, Marilyn, the dog, the lawyers, and the parade of witnesses, they had a wing to themselves and so the hotel just let it go. It wasn't every day you got such big long-term bookings.

Air conditioning in Orlando was an absolute necessity in the September heat. So were lightning rods, as Gould noticed when he started looking around in his way of being curious about things. Asking, he learned the flat palmetto scrub surrounding Orlando and its many lakes was one of the most lightning prone areas in the United States. The federal complex where the courthouse was located was an air-conditioned cab ride away from the hotel. It fronted on a street that paralleled Interstate 4 where it split downtown Orlando like a peach, one half from the other. Outside the air-conditioned courthouse, heat waves rose from the pavement.

By the morning of September 16, the jury was seated and everybody was ready to go.

■■

"An amazing magnificent invention of man. A laser."

It didn't take Warren Goodrich long to get it going. Right from the start, he let the jury know that this was no small potatoes case they were sitting on, but was a case that should resonate in their imaginations. "A machine which has changed our lives. Changed our industry, and changed our lives."

There was a sense of inevitability now. Gould could feel it as he listened to Goodrich tell the jury what the case was all about, the kind of feeling when it was too late to turn the canoe around and paddle upstream from the waterfall. This was it, the main event. A pink certificate attesting that the '845 optically pumped amplifier patent had survived the rigors of reexamination resided in the law firm's files in Westfield. The patent had been tested in interference, re-examination, and in the non-jury trial before Judge Conti. Now a jury would get its crack at Gould. Gould felt a little tightening of his nether parts. Dozens of laser companies had agreed to be bound by the outcome. Millions upon millions of dollars rode on the events of the next few weeks.

Goodrich was amazing, he really was. Gould had the greatest admiration for these lawyers who could enter a case knowing nothing whatsoever about technology, and end up being able to explain it off the tops of their heads in the most accessible terms. Which was what Goodrich was doing. He sketched Gould's history, his invention, and gave a serviceable explanation of quantum mechanics. Then he barreled forward to invoke the Constitution, "a short document for a great nation," and its provision for patents. "Most civilized nations recognize them because what they do is they encourage people to advance science and industry and technology," he said. That was a nice touch, inspirational and subtle all at once.

He spun a little yarn about an automatic hat tipper, invented so men could keep their hands in their pockets on cold days, to explain that some patented inventions were more useful than others. Now he was explaining how Gould's amplifier differed from an oscillator, and he certainly wasn't belaboring the

technology in spite of the fact that two of the jurors were engineers who pre-
sumably could grasp it. No, Goodrich was saying that the amplifier was like a
chair without a back, that if somebody patented a chair with a back, that still
left the field open to patent the basic device, a seat with four legs that was a
component of every chair that had a back on it. Gould had never thought about
it quite that way.

Then he talked a little bit about the reexamination, in which "hardly in-
significant opponents" including Control Laser filed tons—well "perhaps not
tons, but reams and reams of paper"—to try to show the patent was invalid, at
the end of which the patent was nonetheless upheld.

The evidence would show, he told the jury at the end of his oration, that not
only did Gould deserve his patent, but Control Laser had infringed it.

Control Laser's Duckworth wasn't quite the wordsmith Goodrich was. He
didn't have the same flair for the apt metaphor, the sweeping example, or even
the lasting impression. "In case anybody has forgotten, I am Robert Duckworth," he reminded the jury, and went on to argue points that others had argued in attacking Gould's patent—that TRG hadn't been first to build a laser,
that Gould's notebooks were flawed and incomplete, that Gould had cribbed
from the preprint of Schawlow and Townes's *Physical Review* paper, and that
his patent application was inadequate.

He did some hand wringing, saying that Control Laser had paid royalties to
the Research Corporation on Townes's maser patent, to Bell Labs on the
neodymium YAG laser on which it held a patent, and to Hughes Aircraft on the
ruby laser.

And he told the jury that all of the reviews of Gould's patent to date had
been inadequate, that one of the most exhaustively scrutinized patents in history had still not been scrutinized enough. "We are here to supplement the
record that was incomplete before the Patent Office," Duckworth said. "You
are going to be the first group to hear all of this evidence presented the way
that it's going to be presented."

The judge, the jury, and both sides hunkered down for a long trial.

■

Gould wasn't even allowed to pick out his own ties. Goodrich apparently had
made a study, and he said people would react adversely to certain kinds of
neckties and so he'd be happy if Gould would let him pick his ties as well as follow Goodrich's general guidelines for the rest of his attire. He didn't want
Gould wearing striped ties that might make the jury think he was an Ivy Lea-

guer or in some kind of club, or those yellow ties that were popular among the kinds of young urban professionals who were making cigar smoking popular again.

Gould didn't mind. He'd never cared much about clothes. The business of testifying was hard enough as it was. He'd gotten good at it because of the constant repetition, but it still required enormous concentration. Each night at dinner, he and Marilyn ate apart from the lawyers in the hotel restaurant because the lawyers' conversations tended to distract and upset him. He needed to think not about strategies and outcomes, but about his path to the laser.

When Goodrich called him, Gould told himself nothing he had ever done, with the exception of the laser itself, would be as important as the next few days. Ensconcing himself in the blond wood witness box, he summoned all the resources at his command.

And he went through it all again, his lifetime of events, interactions, and discoveries, his desire to invent, his misconceptions about the patent law, the precise timing of what he knew and when he knew it. At first he was up and down, pointing out the differences between radio waves and microwaves and light waves, the basics of optics and the visible spectrum in charts and demonstration tools. Seated, he kept wanting to lean back in his chair, but some genius of a designer had placed the microphone in a fixed position built into the wood surround so that he had to lean forward to be heard.

He talked about his interactions with Townes, his understanding and mild interest in pushing amplification by stimulated emission into the visible range until he woke up that November night in 1957. And then, there it was, the transformation. The sudden insight was something Gould never tired of describing. In all the depositions, all the examinations and cross-examinations, the same thing always happened, the freshness and energy welled up and Gould become something more than an aggrieved patent holder looking for his due. He became an inventor, a discoverer, a pioneer. A schoolchild could look at this grandfatherly man and, seeing him turn young again, know what it was to have your knowledge coalesce in a new way, see the joy of finding yourself on a new plateau where no one else, at just that moment, stood.

"It suddenly flashed into my mind that the amplification of light was going to do something quite remarkable, or could be made to do something remarkable compared to radio waves," he said. "If you made that amplifier a long thin tube, light traveling down that tube and being amplified by these atoms that had been put in the right population inversion, that would form a beam instead of light going in every direction. You could generate a straight beam and concentrate the amplified light power into a beam of great intensity. You could focus that beam down to a tiny spot and get a fantastic intensity. And that thought was exciting. That immediately opened up my mind. My God, you could do things with this which have never been done before!"

As descriptions of the eureka moment go, that one was as good as any. Listening to Gould, you understood what it was that coursed through the mind of an inventor when it opened up and flashed to the horizon and saw what no one had seen before.

■

Gould's testimony covered much familiar ground, and so did the testimony of the witnesses who followed. Peter Franken returned as an expert witness. Paul Rabinowitz was called as a fact witness to talk about his work with Gould at TRG, and his experimental partner, Steve Jacobs, was deposed.

But the X-factor was on the other side, the looming presence of the Nobel laureate, Charles Townes.

Marilyn was scared, and she admitted it. The Nobel carried such prestige. It couldn't help but affect the jury. The rumor in the Gould camp was that Townes had volunteered to testify. Still rankled by Dick Samuel's comments at the General Photonics trial in San Francisco, Townes had waited for years to, as he put it, "set the record straight."

Duckworth called Townes on September 28. At seventy-eight, he was still an imposing figure as he took the witness stand. He stood erect, his bearing patrician, his dress impeccable. His face was stern and his pale eyes unsmiling behind his trifocals, like a man who had been called from important duties in the principal's office to administer discipline to an unruly classroom.

Goodrich offered to concede Townes's qualifications as an expert, but Duckworth wanted to lay them out before the jury. "If it please the court," he said, "I would just as soon go through the process."

With that he took Townes through a description of spectroscopy, his creation of the field of microwave spectroscopy after working in radar during World War II, and his development of the maser. "Then in 1957," Townes said, "I started more intensive work on trying to see if we could extend this down to still shorter wave lengths and that was the origin of the laser."

Duckworth started lobbing softballs. "Dr. Townes, approximately how many scientific papers have you published?"

"A few hundred."

"Have you written any books?"

"Yes, a few."

"Have you been asked to speak at scientific conferences?"

"Yes."

"How many?"

"Some two hundred." Duckworth asked him to list the subjects. Townes reeled off "spectroscopy, electronics, electrical engineering, amplification, clocks, atoms, molecules, lasers, lasers and masers. Astronomy. Generally optics, microwave and optical spectroscopy."

"Have you organized any conferences?"

"Well, I organized the first international conference on quantum electronics, which is the field of lasers and masers."

"Have you received an honorary degree, sir?"

"Yes."

"How many?"

"I think twenty-one."

"Is there one particular award that you have of which you are most proud?"

"Yes, I think that would have to be the Nobel Prize."

"For what did you receive the Nobel Prize, Dr. Townes?"

"That was for work in the field of masers and lasers."

"Dr. Townes, do you still have your Nobel Prize certificate?"

"No."

"What did you do with it, sir?"

"The state museum in my home state of South Carolina named a wing of the museum for me and were eager to have that certificate, so I gave it to them."

"How many other awards besides the Nobel Prize have you won, sir?"

"I suppose thirty, something like that."

After this litany, Duckworth introduced Townes's curriculum vitae into evidence and walked him through it. This required the listing of still more awards and what they were for, emphasizing all the while Townes' background in lasers and masers but including NASA's Distinguished Public Service Medal, the National Medal of Science, and even the Southern Baptist Seminary's Churchman of the Year award. At last Duckworth said, "We offer Dr. Townes as an expert in lasers and masers."

"No objection," Goodrich said dryly.

46

Townes recounted the frenzy of work at Columbia in the mid-1950s, the maser, then the push toward amplifying shorter wavelengths, always staking his and Schawlow's claim. When Gould's TRG proposal crossed his desk as an advisor to the government he was taken aback and could not believe that Gould had reached his ideas without help. That fact that there were similarities to what he and Schawlow had written for *Physical Review* could mean only one thing.

"I believed that possibly he had seen our paper," Townes testified. "Normal courtesy would have been to refer to a paper. I was a little annoyed."

Gould had seen the paper, as he had already testified. There were similarities, and he had testified to that as well. But there were far more differences. Nevertheless, the Schawlow and Townes preprint was on the table again as an issue, and along with it the implication that Gould must have stolen their ideas.

Duckworth helped the impression along, questioning Townes more about the article. It was a natural line of attack. The article had, after all, long since achieved iconic status among scientists as the seminal work about lasers. It was the point from which all laser science had taken off, the Declaration of Independence and the Magna Carta rolled into one, a sacred, infallible document. Gould must have started there, too. There was no other way to look at it.

Duckworth asked Townes to compare Gould's patent with the article. "Do you have an opinion as to whether that patent provides any more information to one ordinarily skilled in the laser field on or about April 6, 1959, than did your article as to how to go about the practical implementation of an operable laser?"

"My opinion is no."

"Would that be the same if I left the word 'practical' out and said 'operable laser of any kind'?"

"Yes, that would be the same."

There you had it. Gould couldn't have invented the laser, or the amplifier on which he held the patent, because Townes said so.

■■

Duckworth must have seen what would happen in the cross-examination. He may not have seen every ominous possibility, but he certainly saw one of them. Before Goodrich started his questioning of Townes, Duckworth asked the judge for an order "to prevent Mr. Gould from attempting to prove that the Schawlow-Townes patent discloses an inoperative device, or that the disclosure of the Townes' patent is not enabled."

This would have saved Townes the embarrassment of revisiting the Gould versus Hellwarth interference ruling, the decision that said the laser Schawlow and Townes described wouldn't have worked. Judge Fawsett deferred a ruling, and Goodrich took his turn.

The snowy-haired lawyer got right to the point, taking aim at Townes's impartiality. "The history of science is filled, is it not, with examples of scientific truth being held back because of the bias of those, of other eminent men of science in the field who simply were not ready to accept some different idea, isn't that correct?"

"There are some such cases."

"And just as scientists should try not to inhibit the scientific discoveries of a Pasteur, also witnesses, whether scientific or not, should be willing to concede anything that's in their background which might affect their testimony and let that be revealed to the jury. Do you understand the wisdom and necessity of that?"

"Sure."

"Doctor, it is a correct statement, is it not, that an event occurred some years ago which has affected your feeling about this patent and about this suit and about Gordon Gould, and given you a personal view of this case."

Goodrich reminded Townes of the General Photonics trial in February 1982, and the article in the San Francisco newspaper that suggested that Townes had taken material from Gould.

"I recall. It seems to me Mr. Samuel or some lawyer said that."

"You were not happy with that, were you?"

"I was annoyed."

"And you were troubled by it?"

"A false statement about you is always a little annoying."

"It troubled you, did it not?"

"Troubled me? Well, I say I was annoyed. I don't know how many words you want to use."

"I want to use the word 'troubled.' " Townes had in fact said he was "troubled" by the article when he was deposed less than two days earlier. He recalled the statement when Goodrich read it back to him, and said, under continued questioning, that it was after that that he had run into Duckworth and agreed to testify.

Goodrich wasn't through, not by a long shot. "When you agreed to come and testify, you hadn't read the patent, but you thought you knew a good deal about it, right?"

"Yes, that's quite right, I knew a good deal about it."

What Townes knew, however, came not from the patent itself, but from what Duckworth had told him, what he had read in the news media, what he had garnered from the reports that came to him as a result of the single share of Patlex stock he had acquired for that very reason, what AT&T's patent lawyer, Samuel Dworetsky, who had compiled AT&T's reexamination request on the Gould patent, told him, and from various claims in other Gould patents. "Is that a fair summary of all the information you had?" Goodrich asked.

"I certainly would include my general background and experience in the field, particularly the first patent case we won against Gould, talking with friends about it, including Mr. Dworetsky in particular, who kept me posted on some of these issues," Townes said.

He had also, it turned out, reviewed Harvey Behrend's statements and conclusions, an uncomfortable admission that it took Goodrich more than a few questions to elicit.

"I believe I did see some discussion by the examiner," Townes said finally.

"Well, as a matter of fact, you were reading it to see what the examiner said, weren't you?"

"Yes."

"Why did you want to know that?"

"I never exclude myself from information about something like this."

"Were you told that the examiner was flatly reversed?"

By now it was abundantly clear that science and law were at wide variance over the courtesies a Nobel laureate deserved. Goodrich asked Townes if he thought the trial was a form of appeal from the Patent Office's final and unappealable decision upholding the Gould patent. He asked Townes if he had let the various legal opinions he had read and heard affect his scientific judgment. He revealed Townes's statement in deposition that he hoped Control Laser was going to win. And he worried Townes about his concern about his place in history.

"You have a personal feeling that your patent covered everything, and that Gould shouldn't succeed on his, isn't that right?" Goodrich asked.

"It still does cover everything. What we are discussing here is an improvment patent falling under my original patent. I don't see that there's any problem."

"I am asking you whether it affects you in your thinking that it seems to take a little of the historical importance of your [maser] patent away if this patent is enforced."

"No. I think that history is all well known and has been for some time."

"And do you have the same feeling about the Schawlow-Townes patent?"

"Yes. The more important lasers become, the more important my patents are. So I am glad the laser field will grow and prosper, and additional improvements, that's great with me."

"Is that part of why you are here today, then, you want to see these other patents approved?"

"I would like to see these patents judged on the basis of good scientific information, that's why I am here."

Gould didn't discount the animus Townes felt toward his claims. But he couldn't help but feel a little sympathy for the other man as the questioning continued. He was glad he had never had to face a cross-examination like the one Townes was facing. Maybe he had never testified with as much baggage as Townes apparently brought to court. Still, he wouldn't have blamed Townes if he had wanted to go home and stick pins into a lawyer doll. Which would have been completely unscientific.

Goodrich was unrelenting. Now he was asking Townes about examiner Nelson Moskowitz's opinion that laser amplification had been achieved before oscillation, which would make the oscillator an improvement of the amplifier. Townes said he simply didn't believe it.

"What you are saying is that you have a closed mind on the subject," Goodrich said. "Your only opinion is that without knowing the facts, without knowing the circumstances, without knowing who was saying who did what when, without knowing any of that, you simply have a scientific opinion that it was impossible for laser amplification to have occurred before laser oscillation. That's what you are saying?"

"I never said it was impossible."

"You say 'never has happened.' "

"That's right. The record indicates that it hasn't happened."

"How do you know it never happened?"

"I know a great deal about laser physics, and all the other articles which have been published. If anybody can point out something different to me, why, I would be happy to change my opinion."

"I wouldn't ask you. I am asking, you do have this firm opinion that laser amplification, as you define the term, could not occur before oscillation, is that right?"

"I didn't say that, no."

"Is that true?"

"No."

"It could happen. Is that right?"

"It could, yes. It's possible."

■■

Goodrich made point after point that weakened Townes's testimony. He forced admissions that removing the mirrors and closing the end of an oscillator created an amplifier, and that Townes had not thought of the twin mirror solution that the Fabry-Perot interferometer represented until Schawlow hit on it. He pursued Townes over listing anhydrous europium chloride as a potential laser substance when Pringsheim, the authoritative reference on the matter, had shown it would not fluoresce and therefore could not lase.

Townes insisted through question after question that somebody skilled in the art, trying to build a laser, would look further based on the fact that he and Schawlow said it would work. "We had more recent information," he said, as if the fluorescence of matter was something that would change with time.

"You think that that is the way that you go about teaching and disclosing how to build and make a laser?" Goodrich asked.

"Yes. Yes."

"All right."

"I'm sure there are other sources that have some errors in them, but there are some sources which are correct, too," said Townes.

"And there are some disclosures made that turn out to be incorrect, are there not?"

"Surely."

That was as close as Goodrich came to revealing the fatal flaws in the Schawlow and Townes patent, but by then he didn't really need to.

■

To Gould's thinking, Goodrich's most telling point—among many—was one that turned the tables on the defense's implication that Gould had lifted his ideas from Townes. Goodrich started by asking Townes about his original maser patent application.

"Do you recall that on May 6, 1955, you applied for your maser patent?"

"I think that's correct."

"Does this patent application contain a specific suggestion of how to pump the molecules or atoms in order to achieve a population inversion?"

Townes responded that the application "contains a number of suggestions of that type." But as Goodrich pursued the question, Townes admitted they were only suggestions.

"Doctor, I didn't ask you if there was any mention in your application of optical excitation. I asked you if there was any mention of a means of pumping."

"That is not detailed here. Not in this exhibit you have given me."

"Not in your patent application. All right."

"Not in this particular one, no."

"This is your patent application, is it not?"

"Yes, I guess that was the initial application, I guess that's what you want to call it."

"Now, do you have a copy of your patent that was issued to you on March 24, 1959? This patent was issued on your original application, as it was amended by continuation application that you filed on January 28, 1958, is that correct, sir?"

"I think that was one of the continuations."

"And with that continuation then, this patent issued to you on March 24, 1959."

"Yes."

"Would you look please in that patent which issued after your amendment, and see if in there you find any specified method for obtaining the population inversion. Does that describe then a method of creating a population inversion that's in your patent as amended?"

"Yes."

Goodrich had Townes read from the patent. His next question was, "That particular description is a verbal description of what in fact is the pumping scheme shown on plaintiff's exhibit 13, is it not?"

Townes had already identified the exhibit, actually two diagrams, as representing Gould's description of optical pumping in the notebook Townes had witnessed on January 3, 1957. He did his best to say his amendment hadn't come from Gould. "Well, I would say that was a verbal description of Prokhorov's system, which was published."

"Move to strike as not responsive," Goodrich said. He was determined to get Townes's admission that regardless of the source he chose to cite, the Gould notebook also contained a description of precisely the scheme of optical pumping that Townes later added to his maser patent—the same patent that Townes had later used to license laser manufacturers.

"That is identical with that, yes," Townes said finally.

47

The trial seemed anticlimactic after Townes left the courtroom and returned to California. Gould felt that Townes's testimony hadn't hurt him. Goodrich had painted the professor as a biased witness, planted the suggestion that Townes might have taken optical pumping for a maser from Gould's notebook, and laid out charts that showed there were actually few similarities between the Townes and Schawlow optical maser paper and Gould's laser proposal. Gould would never deny that Townes was a great scientist, but he had lost some of his mystique on the witness stand.

Rosh Hashanah and Yom Kippur, doctor's appointments, and other matters interceded to lengthen the proceedings. Duckworth summoned a parade of other witnesses, including the one who asserted that optically pumped lasing, triggered by the sun, occurred in the atmosphere of Mars.

Goodrich made short work of him. "Doctor, you haven't discovered any evidence at all that there are little men up on Mars that have made a laser?" he asked.

"No, sir."

"No evidence that there are Russian cosmonauts up there that are doing it, have you?"

"No, sir."

"So we're not talking here about any kind of mechanical device, we're talking about something up there in the atmosphere of Mars, is that right?"

"That's correct. There are no mirrors in orbit providing an oscillator, a resonant cavity."

"We're talking about an atmospheric phenomenon."

"That's correct."

"Can I, in conclusion, assume that as far as scientifically possible, you've excluded the possibility of this being Darth Vader and the Empire striking back up there?"

"Oh, I think so, yes."

Marilyn had her own dress code to conform to, this imposed by Goodrich's secretary who attired Marilyn in sober working women's outfits composed of silk blouses, and skirts and jackets in dark colors that showed she was serious and respectful of the court. Her demeanor was highlighted when the wife of the Control Laser chairman, van Roijen, dropped in from time to time, apparently straight from the country club where she had not taken time to change out of her golf skirt and tennis shoes. Well, that was the way people dressed in the vacation land of Orlando. Van Roijen's aviator sunglasses were part of the same deal. You didn't come to Orlando to dress up. Until you remembered that what was going on here was a company with one hundred and twenty employees that did $15 million a year in laser sales and was charged with infringing Gould's patent for ten years. The back royalties, at the 5 percent rate Gould and Patlex were demanding, came to almost $3 million—20 percent of Control Laser's annual gross. To say nothing of the company's heavy legal fees. No, this trial was no afternoon by the pool.

On the flip side, Marilyn had no trouble remembering that the man she had been with for almost twenty-five years was on the threshhold of something he'd been fighting for since before she'd met him. She could endure a suit and silk blouse and one of those floppy silk bow ties if that was what it took to get them across the finish line.

■■

On October 22, the trial reached final arguments. The courtroom had gradually been filling. Day to day, more spectators joined the phantom jurors as a conclusion neared. Local and trade magazine reporters, attorneys representing other laser companies, people close to Control Laser in Orlando filled the spectator benches and tested the capacity of the small courtroom. Gould wasn't the only one who seemed to be holding his breath.

Control Laser's case had been composed almost entirely of the persistent suggestion that Gould had taken his ideas from Townes because the timing of his proposal had coincided with the Schawlow and Townes article. Goodrich attacked it head on. "Coincidences don't count," he said. "We know that

Townes was shown the optical maser by Gould one year before he amended his application. Do we come in here and say he stole that from Gordon Gould? We do not. It may be that Townes forgot where he got it. It may be he derived it from him. Maybe he didn't. Coincidences do not prove derivation.

"I do not know if counsel will continue with the suggestion that all of this [the information in Gould's patent] was derived from Dr. Townes. I hope not. It's unseemly for an infringing defendant, whose principal witness, if you're going to talk about circumstantial evidence, looks more like he took something. And I don't say that he did. I say that's why you should go by solid evidence.

"All of the evidence shows that Gordon Gould invented every single invention set forth in these broad claims. And had it witnessed and corroborated and worked along with his lawyer until the patent application was filed. He derived it from no one."

"It's clear that Gordon Gould's ideas were coming from Charles Townes," Duckworth said in his summation. "I'm sorry if it's unseemly. To me the evidence is practically irrefutable."

More than that was said on both sides, but what it came down to after almost thirty years was that the arguments hadn't changed. After Goodrich rebutted, saying that "the time has now come for Gordon Gould's inventions from many years ago to receive their recognition and to be enforced," Judge Fawsett started reading her instructions to the jury.

■

The weekend arrived without a verdict. Gould and Marilyn rented a car and drove to Daytona Beach with Angie. By the end of the following Tuesday, there were stirrings that the jurors had made up their minds. Before their decision could be announced, the judge had to make a trip out of town.

This made Tuesday and Wednesday nights unbearable to Gould and Marilyn. Marilyn was suffering anyway, due to an unfortunate decision to hang on to the leash after Angie bolted after a seagull on the beach. She received a dislocated shoulder and could only sleep on one side. But she struggled into another businesswoman's suit, and Gould into his prescribed attire, and they were in court at eight minutes after nine the next morning when the jury filed into the courtroom and the foreman, whose name was Threadgill, handed the court clerk a thick envelope that contained the verdict.

The verdict forms were complicated. Control Laser made several different lasers, and the jury was required to judge the case for infringement on each one. In order to reach those questions, it had had to decide in a variety of ways whether Gould's patent was valid, and then to decide whether Control Laser had infringed it by making that particular laser.

The litany of rulings began.

Not for half an hour did Gould glean the direction of the wind. Some of the rulings seemed to go against him, some for. Ever so slowly, it dawned on him that what he was hearing was good, all good. He felt relief. Jubilation surged against his weariness, but he cautioned himself. Thirty years had taught him to do that. The readings continued. They seemed to go on forever. He looked at Dick Samuel, and received a nod of assurance in return. Goodrich was looking satisfied, while to one side the defense was sending out signals of gloom. From behind him, Gould started to receive vibrations of joy from Marilyn.

"Question twenty-three," the jury foreman recited. It was the last important question. "Have defendants proven by clear and convincing evidence that any of the twelve claims of the '845 patent do not distinctly point out and claim an invention?

"No."

"Does that represent the unanimous verdict of all jurors in this case?" Judge Fawsett asked.

"Yes, ma'am, it does."

Gould had a little smile by now, a quiet, tired kind of smile such as a marathon runner might have when he crosses the finish line and thinks that while maybe he has a few yards left in him he's gone far enough and he's glad, really glad the race is over. A few minutes later, with van Roijen's wife sobbing and van Roijen inscrutable behind his sunglasses, Gould and Marilyn were hugging in the courtroom.

■■

The jury wasn't finished. It had to come back the following Monday for a second trial to determine damages. Judge Fawsett promised it would be short.

Control Laser made a desperate countermove, suing Gould for fraud and charging that he lied to obtain his patent. The judge dismissed the countersuit, and before the trial for damages could begin van Roijen looked at his options. After ten years of legal games, delays, and the unauthorized use of Gould's patent by Control Laser, Patlex was in no mood to be patient. The laser industry had been so intransigent for so long that Dick Samuel wanted to set an example that would bring the rest of the companies in line. The message was simple: You fight with us, you lose your company. He told van Roijen that he was welcome to take a chance on the jury's setting a lower royalty rate and damages than Patlex wanted. But if he did, the only way Control Laser would get a license to use Gould's patent—and to stay in business—was by paying the $3 million in back royalties at the 5 percent rate.

It was money the company didn't have. Control Laser could have entered Chapter 11 bankruptcy to continue operating while it appealed the case, but the safer route, and the better one for its employees and stockholders was to turn the company over to Patlex. Van Roijen gave up. He and his board resigned and handed the company to Patlex.

Patlex's stock already had bounced from $7.75 to $11.25 on the day the verdict was announced, while Control Laser had lost almost a third of its value, falling from $2.75 to $2. The takeover by Patlex sent Control Laser up again, with investors believing the company would find new stability.

"Is that it?" Gould asked Samuel when the negotiations were over.

"You know, Gordon," Samuel said, "hard as it is to believe, I think it is. I think that's it. I think we'll sign up the other companies, and the royalties will start coming in at last. Did you ever think it was going to take this long?"

"No," Gould said, "and I'm glad I didn't."

In fact, more waiting was ahead. But the hardest part, the wondering, was over.

48

Gould and Marilyn were still in Orlando on Tuesday, November 3, when the next piece of news arrived. This time, it wasn't unexpected. The Patent Office, with the order from Judge Flannery finally authorizing rather than directing it to do so, issued Patent Number 4,704,583, for Gould's gas discharge laser amplifier using collisions of the second kind.

"It's actually issued," Gould said as they packed and prepared to head home with Angie to Bonum Creek, Gammon, the crab traps, and the winter wheat. "I'm amazed."

His other comment was that the Patent Office "creaks."

It was just ten days short of the thirtieth anniversary of the notary's seal on Gould's first notebook. The gas discharge amplifier and the optically pumped amplifier patents covered 80 percent of the lasers made in the United States. The market for those lasers had grown to $500 million a year, and was doubling every four years. The optically pumped amplifier patent had just seven years left to run, but the gas discharge amplifier patent would cover the lasers that used it for a full seventeen-year patent term. The long delays forced on Gould by the Patent Office guaranteed to make him a multimillionaire.

Good news followed good news. It was unstaunchable. It welled up like water from a mountain spring or, as Peter Franken had impishly described the burst of light from the Q-switch in one of his many court appearances, like "a giant orgiastic burp."

On December 15, U.S. District Judge George Revercomb in Washington agreed with Gould's attorneys that the Patent Office had exceeded the scope of the reexamination statute in trying to take away Gould's patent on the uses of lasers. The law said newly asserted prior art—patents or printed publications—was the only thing that could be considered. "Consequently," the judge wrote, "the Commissioner may not on reexamination consider whether the specification of a patent . . . contains an enabling disclosure." In other words, refusing to certify a previously issued patent on grounds of insufficient disclosure broke the rules.

On April 26, 1988, following Judge Revercomb's authorization to do so, the Patent Office issued its pink reexamination certificate of patentability for Patent No. 4,161,436, Gould's use patent.

A month later, on May 24, the PTO issued Gould's Brewster's angle window patent, Patent No. 4,746,201. This, too, had required a struggle although it was, compared with the others, a relatively minor patent. Behrend, of course, had rejected it. He had been upheld within the Patent Office, and Wepner had taken the case to the Court of Customs and Patent Appeals. But by now the Patent Office had seen that the tide was running in favor of Gould's patents, and its solicitors decided that further litigation was senseless. The PTO asked the court to remand the case to its Board of Appeals, and on the second go-round the board reversed all of Behrend's original rejections.

Gould now had four basic laser patents that had been tested in every imaginable way. They were a fraction of his total of forty-eight patents overall, accumulated at TRG, Brooklyn Polytechnic, and Optelecom, but they were by far the most important in terms of both licensing income and the emergence of Gould's reputation as the real inventor of the laser.

Patlex had announced on the eve of the Control Laser trial that it had signed licensing agreements with major companies including Eastman Kodak, Amdahl, Chrysler, EverReady Battery, and Union Carbide. The victory in Orlando brought more of the two hundred laser manufacturers and users to the table, taking licenses largely on Patlex's terms. Back royalty payments ballooned the Patlex treasury.

In September 1988, when Spectra-Physics, the world's largest laser manufacturer, joined number two Coherent in licensing Gould's patents, well over a third of the companies subject to royalties had signed agreements and the pace of licensing increased.

The Gould laser patents eventually produced $12 million a year, divided among Patlex, Gould, and Refac.

■■

All four of Gould's basic laser patents were still in force in 1991, when Gould was inducted into the National Inventors Hall of Fame in Akron, Ohio, where he joined, among others, Charles Townes. The hall, a joint project of the Patent and Trademark Office and the National Council of Property Law Associations, was founded in 1973. Its first member, Gould's hero Edison, was named alone in the hall's inaugural year to signify his Olympian stature among inventors. Gould's citation credits him with the invention of the two vital amplifiers, and with coining the name "laser."

Gould's first notebook, in which he named the laser from its acronym and laid out the opposing mirrors that became the key to the by-now-familiar laser beam, already resided in the collection of the Smithsonian Institution.

Gould finally gave up his three-pack-a-day cigarette habit in 1991, along with Marilyn. They were married a year later on his birthday, July 17, making official what had long been the fact of their relationship.

Gould gave away some of his inventor's wealth a few years later, donating $1.5 million to endow a physics professorship at his alma mater, Union College, in 1995. The post bore Gould's name; the appointee would be the R. Gordon Gould Professor of Physics. But Gould, in making the donation, stressed that he had done it to honor his old professor, Frank Studer, who had taught him to love light and the science of optics.

By then, Gould's hard-won optically pumped amplifier patent had expired, in 1994. His patent on the uses of lasers, its prophecy long since fulfilled, expired in 1996. The gas discharge amplifier patent remains in effect until November 2004, and the Brewster's angle window patent until May 2005.

"I hope I'm still around," Gould said.

■■

In June 1995, the United States adopted a new patent law. "Harmonization" was the key word, in that it brought America into conformity with the patent laws of other industrialized nations under the General Agreement on Tariffs and Trade. A key aspect of the law was a new patent term of twenty years, but a term that began on the date of application rather than the date the patent issued.

Had such a term applied to Gould, his optically pumped amplifier patent would have protected him against infringement for two years, and his use patent for one.

The big international corporations were all for the new term, and for further harmonization with the patent laws of other countries, especially the publication of patent applications rather than keeping them secret until a patent issues.

This, they said, would assure the quick entry of new technologies into the economy and keep unscrupulous inventors from "gaming the system" with "submarine" applications that lurk in pendency for years until patents issue that can be applied to mature industries. Nothing, they said, would prevent the licensing of pending patents. Nobody mentioned Gould's name, at least not officially.

Independent inventors rose up in protest.

Both sides had a chance to air their views in hearings in November 1995. U.S. Representative Carlos Moorhead of California had proposed adjustments to the law, adding provisions that would extend the patent term in cases of interferences and other Patent Office delays but at the same time publish applications eighteen months after they are filed. Another bill advocated returning the United States to the old way of doing things, with a seventeen-year patent term from the date the patent issued and applications that were kept secret until then. Armies of corporate patent lawyers marched on Capitol Hill, and so did inventors, armed with the Constitution and the argument that, as Abraham Lincoln put it, the traditional patent system added the fuel of economic incentive to "the fire of inventiveness."

Ultimately, both sides were successful. The American Inventors Protection Act of 1999 keeps the patent term at twenty years from the date of application, but extends the term for applications pending for more than three years due to interferences, secrecy orders, appellate reviews, and "unusual delays" by the Patent Office. Diligent applicants under the new law are guaranteed a minimum seventeen-year patent term.

An inventor in the United States who files applications both at home and abroad will have his application published after eighteen months, which corporations say will help them stay abreast of new technology. Applications filed in the United States alone, however, will remain secret until the patent issues.

Under the U.S. patent law, the United States remains the only country in the world that awards patents to the inventor who can prove he was first with a new invention, rather than the one who was first to reach the Patent Office with an application.

■■

The twentieth century ended with a flurry of list-making. There were lists of biggests, and bests, and favorites of the past hundred years. Among them were the century's most significant inventions, and no such list overlooked the laser.

In the more than forty-two years since his flash of insight, Gould had seen the laser fulfill all of his predictions and more. Lasers had found vital uses in medicine, communications, manufacturing, and materials processing. They measured distances to within millimeters, and a laser beam bounced off a retroreflector placed on the moon by Apollo astronauts tracked minute shifts in the San Andreas fault. Lecturers used them as pointers, hunters and snipers used them to fix targets, schoolchildren used them to make mischief. Lasers were used in games of tag, and in military training. Compact disc players and other consumer products that relied on them were ubiquitous, and hardly any store or supermarket that passed goods across its checkout counters was without them. They made possible vast amounts of data storage and its instant retrieval. Displays and reprographic systems used them. They guided missiles, among their many military and aerospace applications. Generals still dreamed their dreams of lasers knocking missiles from the sky, while researchers continued to seek ways of using lasers to trigger hydrogen fusion to create an unlimited source of power generation.

Some $200 billion worth of lasers and laser systems had been sold by century's end.

The beginning of the new century found Gould and Marilyn, after commuting back and forth from Bonum Creek for several years, living in a ski village in the mountains of Colorado, and back in New York, where they have an apartment close to Lincoln Center. Gould has some entrepreneurial interests, and his most recent foray into invention is a device that uses ultrasound waves to loosen crowns cemented to the roots of teeth, so that dentists will no longer have to whack away with hammers to get at the underlying roots to do their work. It was entering clinical trials, a preliminary step toward federal Food and Drug Administration approval, in the spring of 2000.

While he waits for the results, Gould enjoys walks in the mountains with Marilyn and their new dog, Omi, a mix of malamute, husky, and wolf that thrives in the snow and whose name is short for Naomi. Marilyn sings in the Colorado Symphony chorus, and she and Gould sponsor chamber music concerts at their spacious home to benefit the Breckenridge Music Institute, which has a chamber orchestra and puts on a summer music festival. They are patrons, in a minor way, of artists and sculptors whose work they love. Gould no longer skis, but he has a pair of snowshoes hanging by the garage door, and even when the snow has drifted higher than the window sills he swims fifty laps each morning in the heated indoor pool he purifies not with chlorine, but a system using ultraviolet light.

Gould stays away from Washington for the most part. He refrained from the furor that surrounded the changes in the patent law. The father of the laser turned eighty in July 2000, and is quite content, he says, to be retired from the patent wars.

ACKNOWLEDGMENTS

The research and preparation of this book involved dozens of interviews, hundreds of shorter conversations, and countless hours reading depositions, trial transcripts, and court decisions, as well as records of Patent and Trademark Office proceedings, FBI and Defense Department files, and personal archives. As a new arrival to both physics and patent law, I suspect that many of my inquiries were exasperating in their need to reduce complicated concepts and processes to utter simplicity. But they were necessary for both my understanding and that of readers who, like me, find the physical and legal worlds fascinating and capable of producing a hell of a good story, but in need of occasional deciphering. I thank one and all for bearing with me. Responsibility for any shortcomings in interpretation rests solely with me, and not my sources.

My first and most profound thanks go to Gordon Gould, for his infinite patience, his unflagging candor, and his unfailing sense of humor, and to Marilyn Appel, whose memory is as keen as her perceptions. Without them this book would never have been written.

Close behind them on the scale of gratitude are Steve Jacobs and Paul Rabinowitz, Gould's former scientific colleagues, who shared their memories of and enthusiasm for Gould's work unstintingly. It was through Jacobs's friendship with Simon & Schuster's Alice Mayhew that Gould's amazing story came to my attention.

Patent attorneys Dick Samuel and Bob Keegan also occupy the upper strata of contribution. Their help in navigating Gould's patent history and, in

Samuel's case, the business side of his patent litigation as well, involved an uncommon generosity of time.

Hank Muetterties went beyond the call of duty in retrieving files and court transcripts from dead storage in the summer heat of Las Vegas. Gary Erlbaum explained the origins and many of the business dealings of Patlex. Ken Langone provided insights into the financing.

Joe Littenberg, Roy Wepner, and Bill Mentlik helped me follow the many threads of the patent litigation. Roger Cook recounted courtroom scenes.

Jack Kotik and the late Larry Goldmuntz recalled the origins and early days of TRG. Ted Shultz, Ray Forestieri, Herb Gresser, Ben Senitzky, Gerry Grosof, Steve Barone, Maurice Newstein, Herman Cummins, John Poulos, and Bob Chimenti added recollections of laser work at TRG and elsewhere.

Charles Townes was gracious to recount events at Columbia and at Bell Labs, as were James Gordon and the late Arthur Schawlow. Herwig Kogelnik, Arthur Torsiglieri, and Lou Canepa also shared the Bell Labs point of view.

Bill Bennett was generous in his recollections of his work on the gas discharge laser, as was Don Herriott.

Several people shared their recollections of Columbia University's physics department in the 1940s and 1950s, among them Bill Nierenberg, Martin Perl, Alan Berman, Bob Novik, and the late Peter Franken. Dr. Norman Christ of the physics department, and Bob Nelson of the Columbia information office also were most helpful.

Bill Culver talked about his partnership with Gould at Optelecom.

John Coleman, Howard Turner, Eugene Lang, and Joel Mallin filled in bits of the picture about Gould's business dealings.

Drs. Milton Zaret and Goodwin Breinin talked about the first experiments in laser eye surgery, and Dr. Frank Spelman recounted his laser surgery on Gould.

The information offices at Union College, New York University, and the New York Polytechnic Institute were helpful in tracing names and providing background.

The files of the City University of New York, the Federal Bureau of Investigation, the Air Force, and the papers of Adam Yarmolinsky at the John F. Kennedy Library in Boston yielded the fruits of Gould's political history. Joseph Prenski provided recollections of that era.

Heartfelt thanks for his forbearance goes to my editor, Bob Bender, for whom my time spent awaiting files under the Freedom of Information Act was surely not good news. My agent, Lynn Nesbit, has been patient, too, and I owe her thanks for that and more.

Finally, to Barbara, you are the reason this is all worthwhile.

INDEX